THE
MISMEASURE
OF
MINDS

STUDIES IN SOCIAL MEDICINE

Allan M. Brandt, Larry R. Churchill,
and Jonathan Oberlander, editors

This series publishes books at the intersection of
medicine, health, and society that further our
understanding of how medicine and society shape
one another historically, politically, and ethically.
The series is grounded in the convictions that
medicine is a social science, that medicine is
humanistic and cultural as well as biological,
and that it should be studied as a social, political,
ethical, and economic force.

THE
MISMEASURE
OF
MINDS

Debating Race
and Intelligence
between *Brown*
and *The Bell Curve*

Michael E. Staub

The University of North Carolina Press
Chapel Hill

Designed by April Leidig

Set in Warnock by Copperline Book Services

Manufactured in the United States of America

The University of North Carolina Press has been a member
of the Green Press Initiative since 2003.

Jacket illustration © iStockphoto.com/Hendra Su.

Library of Congress Cataloging-in-Publication Data

Names: Staub, Michael E., author.

 Title: The mismeasure of minds : debating race and intelligence between
 Brown and The bell curve / Michael E. Staub.

Other titles: Studies in social medicine.

Description: Chapel Hill : The University of North Carolina Press, [2018] |
 Series: Studies in social medicine

Identifiers: LCCN 2018016978| ISBN 9781469643595 (cloth : alk. paper) |
 ISBN 9781469643601 (ebook)

Subjects: LCSH: Intellect—Social aspects—United States. | Intelligence levels—
 Social aspects—United States. | Intelligence tests—Social aspects—United
 States. | Racism—United States—Psychological aspects. | Intellect—Genetic
 aspects. | Eugenics—United States—History. | Educational psychology—United
 States. | School integration—United States. | Segregation in education—United
 States. | Topeka (Kan.). Board of Education—Trials, litigation, etc. | Brown,
 Oliver, 1918–1961—Trials, litigation, etc. | Herrnstein, Richard J. Bell curve.

Classification: LCC BF431.5.U6 S73 2018 | DDC 153.9089—dc23

 LC record available at https://lccn.loc.gov/2018016978

A shorter version of chapter 3 appeared in Michael E. Staub, "The Other Side of the
Brain: The Politics of Split-Brain Research in the 1970s–1980s," *History of Psychology*
19, no. 4 (2016): 259–73 (© 2016 American Psychological Association).

A prior version of chapter 4 appeared as Michael E. Staub, "Controlling Ourselves:
Emotional Intelligence, the Marsh-mallow Test, and the Inheritance of Race,"
American Studies 55 (Spring 2016): 59–80.

Lucy

CONTENTS

Introduction 1

ONE

Plasticity of Intelligence, Education Reform,
and the Disadvantaged Child 17

TWO

Minimal Brain Dysfunction, Ritalin,
and Racial Politics 49

THREE

The Politics of Cerebral Asymmetry
and Racial Difference 79

FOUR

A Racial History of Emotional Intelligence 109

FIVE

Neuroscience, Race, and Intelligence
after *The Bell Curve* 139

Afterword 169

Acknowledgments 173

Notes 175

Index 211

FIGURES

1. Arthur R. Jensen in 1970 5

2. Richard M. Nixon with Daniel Patrick Moynihan in 1970 21

3. Ernest Chambers in 1968 51

4. Mark A. Stewart in 1967 61

5. Photomontage in the *Journal of Learning Disabilities*, 1970 65

6. Advertisement for Ritalin in the
American Journal of Diseases of Children, 1971 68

7. Vint Lawrence cartoon in the *New Republic*, 1997 76

8. Roger W. Sperry around 1970 84

9. Joseph E. Bogen in the mid-1980s 86

10. Illustration by Bogen from 1975 87

11. *New Yorker* cartoon by Lee Lorenz, 1977 104

12. Advertisement for a "Don't Eat the
Marshmallow!" coffee mug 112

13. "Human Beings Are Not Very Easy to Change
after All," *Saturday Review*, 1972 125

14. *From Neurons to Neighborhoods* graph, 2000 154

INTRODUCTION

I N 1954, in the landmark *Brown v. Board of Education* case, the U.S. Supreme Court ruled that de jure racial segregation in educational facilities violated the equal protection clause of the Constitution's Fourteenth Amendment. It was hailed at the time — and has continued to be acclaimed to the present day — as the historic culmination of a decades-long struggle to achieve racial justice under the law, particularly with regard to the often inadequate public schooling available to African American children. *Brown* thus signaled the end of a Jim Crow apartheid regime and represented real progress as it promised to each American "a right which must be made available to all on equal terms."[1]

Brown's legacy, however, has been far more complex and contradictory than a capsule summary may indicate. With *Brown*, the Supreme Court finally reversed *Plessy v. Ferguson*, the 1896 case that affirmed the noxious doctrine of "separate but equal." Yet it is also true that the *Brown* decision managed to misidentify a key element of the *Plessy* verdict. By the 1950s — as historian Daryl Michael Scott has observed — it was largely taken to be common sense (as it was by Chief Justice Earl Warren, who wrote *Brown*'s unanimous opinion) that the court in Homer A. Plessy's day had been centrally "concerned about Plessy's inner world and that it had simply lacked the scientific knowledge to decide on the plaintiff's behalf."[2] An understanding that science had the ability to analyze an individual's "inner world" became central to the *Brown* opinion; this had not been the concern of *Plessy*. The earlier decision had observed that if African Americans declared white supremacist laws to confer upon them a lower caste status, they had only themselves to blame. "We consider the underlying fallacy of the plaintiff's argument to consist in the assumption that the enforced separation of the two races stamps the colored race with a badge of inferiority," the majority opinion in *Plessy* stated. "If this be so, it is not by reason of anything found in the act, but solely because the colored race chooses to put that construction upon it."[3] But counsel for Homer Plessy had argued that the plaintiff, a Creole from New Orleans light-skinned enough to pass as white, had been denied his *property rights* as a result of de jure segregation because whiteness itself should be recognized as a form of property. Albion Tourgée, one of Plessy's attorneys, had asked the court, "How much would it be *worth* to a young man entering upon the practice of law, to be regarded as a *white* man rather than a colored one?" Tourgée added, "Probably most white persons if given a choice, would prefer death to life in the United States *as colored persons*. Under these conditions, is it possible to conclude that *the reputation of being white* is not property?"[4] Yet the Supreme Court in *Plessy* remained unimpressed with Tourgée's argumentation, deciding with astonishing perversity that the plaintiff's case had no merit because — as a person legally classified as nonwhite — he could not legitimately claim to have been denied the tangible benefits of whiteness.[5]

In overturning *Plessy*, the *Brown* decision in 1954 set aside all discussion of the "tangible" benefits of whiteness and focused instead on the "intangible" damages of blackness. *Brown* rejected the distorted logic of the late nineteenth-century decision, but it did so by insisting that de jure segregation *did* mark African Americans as inferior. *Brown* argued that the wounds caused by discriminatory laws were inescapable precisely because their impairments were *psychological* in nature. *Brown* held that African American children were especially vulnerable to the emotional anguish that resulted from their segregated existence.[6] In shifting from an emphasis on material inequity to a preoccupation with psychological injury, *Brown* turned the causal relationship that *Plessy* had debated on its head. *Plessy* had ruled that a segregated (but otherwise benign) environment did not result in African Americans being denied their constitutional right of equal protection, because separate racial spheres did not do any discernible harm. *Brown* ruled that a segregated environment itself represented a toxic space and that children of color living in that toxic space were pitifully defenseless against its ravages. The postwar liberal consensus in *Brown* thus posited that the malignancy of a segregated environment had of necessity to be eliminated to prevent all further assaults upon the psyches of African American children.

Attorneys for the plaintiffs in *Brown* had put psychology at the center of their legal strategy. At the *Briggs v. Elliott* school desegregation trial in Charleston, South Carolina, in 1951, the first of several federal court challenges to school segregation that had been pursued simultaneously by the National Association for the Advancement of Colored People Legal Defense Fund and that came collectively to be known simply as *Brown*, psychologist Kenneth Clark spoke of the racial preference doll experiments that he and his wife, psychologist Mamie Clark, had conducted with hundreds of young black children.[7] In these racial preference experiments, the Clarks presented African American children with dolls identical in every respect, except skin color. "Show me the doll that you like best or that you'd like to play with," the Clarks inquired. "Show me the doll that looks 'bad.'" And so forth. That these children overwhelmingly ascribed positive qualities to the white dolls and negative attributes to the black dolls meant — at least to the Clarks — that these children suffered from an internalized self-hatred specifically due to their segregated circumstances.[8] In a summary of the doll tests in his testimony at the Charleston trial, Kenneth Clark said that segregated children were "subjected to an obviously inferior status in the society in which they live, [and they] have been definitely harmed in the development of their personalities; ... the signs of instability in their personalities are clear, and I think that every psychologist would accept and interpret these signs as such."[9]

Brown was not the first case to present psychological evidence as part of a legal strategy. "But *Brown* went further," historian Ellen Herman has written. "It illustrated how effectively psychological perspectives on the development of racial identity, and the damages done to it by prejudice, could penetrate the public sphere as constitutional issues."[10] So here in *Brown* the decision turned largely on psychological evidence that (it was said) had proven that racially segregated schools ultimately (and always) undermined the emotional well-being of African American children.

A reliance on psychology to remake law came with its own set of unintended effects. For one thing, a damage hypothesis that cast African Americans as self-hating victims of racial oppression came to dominate (and to haunt) social scientific research and public policy throughout the remainder of the twentieth century.[11] Just as significant, however, was that even as it marked a major turning point in the civil rights struggle, *Brown* inadvertently served to move the national conversation from a discussion of economic and social injustice to a debate over racial intelligence. As historian John P. Jackson Jr., has written, *Brown* encouraged segregationists and their allies to "return to the courtroom with scientific expert witnesses to show that the factual basis that *Brown* rested on was in error."[12] Alert to potential fissures in the Supreme Court decision, segregationists by the early 1960s were arguing that *Brown* had been decided by liberal ideologues who had falsified actual science to serve their pro-black agenda.

Moreover, a renewed focus on the purported correlations between race and intelligence freshly invigorated court challenges to *Brown*. Most notably, the *Stell v. Savannah* case in Georgia saw a segregated school district argue, however disingenuously, that if it could be shown that it had maintained racial separation in its schools, this had been the result of color-blind intelligence testing and had nothing whatsoever to do with racially motivated designs. The Savannah school district brought a string of expert witnesses to testify that to integrate African American children with their lower innate intelligence in (previously all-white) public schools with their higher educational standards "would result in a 40 to 60 percent Negro failure rate in intermediate grades."[13] The judge in *Stell* went on to rule that he had been persuaded by the district's experts and that the segregation of schools could be "attributable in large part to hereditary factors, predictably resulting from a difference in the physiological and psychological characteristics of the two races."[14] Using racist science to battle racial liberalism, in short, the *Stell* case proved to be — as one scholar has argued — a "warped mirror image of *Brown*."[15] But such legal challenges were soon rejected; *Stell* was reversed on appeal to federal court.[16]

More lasting difficulties for *Brown* arrived as a series of publications from psychologists announced that differences in intelligence between the races were insurmountable and far outweighed whatever significance environmental factors might have. There was an essay in 1961 by Henry E. Garrett, retired Columbia University psychologist and former president of the American Psychological Association, in the journal *Perspectives in Biology and Medicine*, in which Garrett coined the term "the equalitarian dogma," defined as the view that "except for environmental differences, all races are fundamentally equal." Garrett baldly concluded that "the equalitarian dogma, at best, represents a sincere if misguided effort to help the Negro by ignoring or even suppressing evidence of his mental and social immaturity. At worst, the equalitarian dogma is the scientific hoax of the century."[17] Psychologist Audrey Shuey's *The Testing of Negro Intelligence*, a tome that ran to more than five hundred pages, ended with an assertion that the data, "all taken together, inevitably point to the presence of native differences between Negroes and whites as determined by intelligence tests."[18] The 1967 collection *Race and Modern Science* included several statements by psychologists and had been

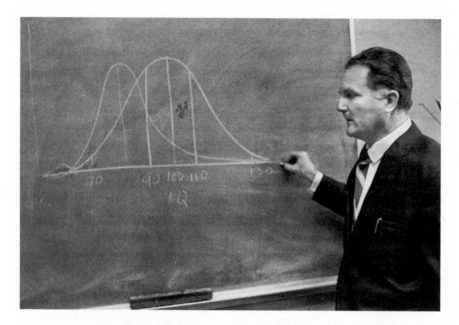

FIGURE 1. In 1970, Professor Arthur R. Jensen used the blackboard to illustrate what he argued was a fifteen-point differential in the intelligence of whites and African Americans. The lower peak represented the average IQ of blacks, while the higher peak represented the average IQ of whites. Jensen—with shockingly blatant but also highly revealing condescension—explained in a profile in *Life* magazine that same year that his views on race and intelligence could actually be thought of as comforting to African Americans. As Jensen said, "As long as people are told there are no differences among groups and individuals, and the only differences are differences society imposes on them, then I think you are going to have frustration, are going to have a kind of paranoia develop, with people wondering, 'Look, these people are making a pretense of giving us every opportunity, but we're still not making it. In what way are they keeping us back?'" Numerous scholars have refuted both Jensen's numbers and his interpretations. Photo by Michael Rougier via Getty Images.

prepared specifically to counter UNESCO's progressive declarations on race and racial prejudice. (One reviewer summarized the collection's import, "This will be a useful source book for racists.")[19] The one-hundred-plus-page article published in 1969 in the prestigious journal *Harvard Educational Review*, written by Berkeley educational psychologist Arthur R. Jensen, simply stated as truth (or perhaps also as dare) that "no one has yet produced any evidence based on a properly controlled study to show that representative samples of Negro and white children can be equalized in intellectual ability through statistical control of environment and education"[20] (see figure 1).

And there was in the *Atlantic* magazine in 1971 Harvard psychologist Richard J. Herrnstein's influential article "I.Q.," which likewise speculated that
"if we make the relevant environment much more uniform (by making it as
good as we can for everyone), then an even larger proportion of the variation
in I.Q. will be attributable to the genes." And it was here that Herrnstein
first floated the concept that an individual's inherited "mental abilities" determined that person's socioeconomic status — a view that would resurface
with far more cultural import in the 1994 book he coauthored with political
scientist Charles A. Murray: *The Bell Curve: Intelligence and Class Structure
in American Life*.[21] If it had ever been anticipated that *Brown* might dis-
mantle once and for all a pseudoscientific view that races (or classes) can be
grouped differentially according to levels of intelligence, it was precisely this
that proved chimerical.

THIS BOOK WILL REVISIT a series of powerfully influential psychological
studies in the postwar decades — along with the conflicting lessons these studies imparted, the debates they provoked, and the policies they encouraged —
that help us to periodize the uneasy and uneven transition between the
Brown v. Board of Education decision in 1954 and the publication of Herrnstein and Murray's *The Bell Curve* precisely four decades later. From one
angle, the book can be read as an investigation of the unfinished paradigmatic shift these two iconic events in the history of postwar psychology
and race still dramatize for us: between an environmentalist and egalitarian perspective that (however problematically) underscored psychic damage
and cultural deprivation and an opposing racialist (if not openly racist) view
rooted firmly in biological determinism. Yet this ever-inconclusive battle
between divergent — yet also mutually imbricated — conceptualizations of
race forms only one crucial background for the alternate history told here.
For I want to stress that to trace debates over race and intelligence between
Brown and *The Bell Curve* is not meant only to indicate a chronological
frame. (The book does at times reach into the prewar era, even as it recurrently documents, close to a quarter century after *The Bell Curve*, that we
still live within the terms of discussion it set.) Instead, my aim is to show
that a history of race and intelligence between *Brown* and *The Bell Curve*
also reconceives what it means to live *in the constant tension between* what
those two paradigmatic text-events stand for: on the one hand, a call for
government policies to alter social environments (based on a conviction that

growing minds are plastic and capable of change), and on the other, an insistence that biological inheritance will inevitably trump most ameliorative efforts and that these efforts are in any event a waste of federal funds. In this way, and additionally, my intention is to foreground that the seemingly irresolvable dispute between environmentalists and hereditarians was never as pure or stable a formation as it has often been imagined.

The book has three further aims. The first is to recover the significance of a sequence of psychological experiments and explorations that ostensibly, at least initially, had nothing to do with race. This research addressed "experimenter bias," "interpersonal expectancy effect," and "learned helplessness," as well as the value of deferred gratification for preschoolers, the symptomatic diagnosis and psychotropic treatment of hyperactivity, the significance of hemispheric differences in the brain, and the noncognitive skills associated with "emotional intelligence." Yet I show how these psychological — and later neuropsychological and neurophysiological — studies nonetheless became inextricably intertwined with public policy discussions of race, racial attitudes, and educational reform.

A second goal is to demonstrate that to study race and intelligence between *Brown* and *The Bell Curve* tells us a great deal about whiteness in postwar America. A striking discovery of my research was the extent to which white members of the middle and upper classes were also addressed and engaged in worries over learning difficulties, beliefs about creativity and intuition, and ideas about the benefits of self-discipline and postponement of gratification. Indeed, and indicatively, race goes in and out of focus throughout the book — as it did in the debates of the last sixty-plus years — for the very elusiveness and allusiveness of the language on all sides is part of the complexity of the topic. Sometimes references to race or class came to stand in for one another, with *middle-class* or *suburban*, for example, becoming code for *white*, while mentions of *impoverished* or *disadvantaged* or *inner-city* gestured to persons of color; a welter of equally unsubtle codes (from *father-absence* to the *"slow" track*) recurred as well; and at other times, a shift in focus from race to class provided the proponents of such a shift with a sly ability to ignore the structures of racism. At yet other moments, however, commentators carefully sorted race from class quite deliberately in their analyses — either puzzling over whether color or low socioeconomic status was the primary determinant of a particular outcome or, for instance, attending to class divisions within racial groupings and noting the commonalities between whites and blacks within a class precisely so as to be able to counter racist assumptions. To invoke color, moreover, did not mean to be racist. On the contrary, such invocations could indicate attentiveness to the particular life-experiences of minority individuals and to the specific consequences of racial prejudice and discrimination, whether coming from (consciously or unconsciously biased) elementary school teachers or from the world at large and the circumscribed limits on life opportunities more generally. Most of all, what becomes clear is that discussions of psychological research and controversies over education and intelligence not only repeatedly provided a means to talk about race without talking about race but also, and simultaneously, had extraordinary consequences for the self-understandings of all Americans.

Last, this book aims to be a contribution to the history of psychology as a discipline and to the history of its growing popularity and social impact in the postwar era. As the chapters unfold, the book traces efforts pursued by many psychologists after *Brown* finally to quell the (always again resurgent) hypothesis that there existed race-based intelligence differentials — even as other psychologists continued to provide data and theories that complicated such efforts. Advancing the causes of the underprivileged, psychologists and allied professionals pursued numerous — often contradictory — stratagems. They argued for the plasticity of intelligence to refute biological determinism, even as they faltered when called upon to produce hard scientific proof that compensatory programming made a critical difference. At other times they conceded that intelligence quotients might be innate and that measurable differences between racial groups might exist but that these differences need not suggest that minority groups were intellectually deficient — rather, they possessed intuitive talents often missing in whites. At yet other times, psychologists advised that resilience and "grit" were — regardless of color — better predictors of life-success than was IQ.

While revealing the continual mutual enmeshment of environmental and biological arguments, then, this book identifies as well an overall historic shift, over the course of the post-*Brown* era. This shift is evident in an uneven but incremental reorientation of focus: from attention to the larger structural harms produced by racial capitalism and toward ever-greater preoccupation with the details of what may be going on inside an individual's brain (whether in the form of sophisticated technologies designed to capture the neurological vulnerabilities of children in poverty or in pervasive injunctions to all Americans to engage in the labor of self-improvement). For gradually, in the context of a wider cultural and professional turn to the neurosciences since the 1990s, we find ourselves increasingly in a realm where seemingly color-blind talk about the individual brain and an individual's behavior, motivation, and responsibility actually serves to mask all manner of historical assumptions about race and class alike.[22] Throughout, however, it becomes apparent that there is a mismatch between the measurement tools used by all sides in the battles over race, class, education, and intelligence and the enormity of the social and political inequities that form the constant backdrop to the debates. At the same time, the book's findings demonstrate, again and again, how extraordinarily influential and relevant psychological experimentation and theorizing has been for both government policy and popular opinion.

The Mismeasure of Minds builds on and benefits from scholarly research in several overlapping areas. It is situated at the intersection of the history of scientific and social scientific attitudes toward minorities and the poor in the twentieth century, including the history of educational reform; the interdisciplinary fields of whiteness studies and critical race theory; the histories of psychology and the neurosciences; and the history of the transfer and adaptation of theories and the interplay of the behavioral sciences and medicine with culture and politics — with a particular focus on the 1950s through the 1990s but reaching also into our twenty-first-century present. The book explores how psychological theories migrate into popular culture and public policy and attends to the ways in which theories and concepts, both when they have merit and when they are manifestly incoherent, can nonetheless be profoundly consequential.

CHAPTER 1 TAKES UP the pressure under which preschool enrichment programs like Project Head Start — promoted by President Lyndon B. Johnson as signature components of his War on Poverty — found themselves needing to demonstrate that they were worthy of investment, rather soon after they had been launched. President Richard M. Nixon expressly invoked psychological research to justify his administration's efforts to question the efficacy of such programs. In this way, he redirected for his own purposes a debate among psychologists that had been ongoing since the later 1950s about how best to address the situation of children who were failing to learn. All sides — for there were more than two — relied on psychological research, and crucially, all sides, liberals and conservatives, concurred that IQ was the appropriate metric for measuring whether particular approaches or programs were succeeding. A most significant turning point came in 1966 with the *Equality of Educational Opportunity* report (better known, after its lead author, as the Coleman Report). Expected to demonstrate that students in segregated schools lagged in IQ scores, the Coleman Report instead produced evidence that some children in predominantly African American schools were doing better than those in desegregated ones. Caught off guard by the data, Coleman attempted to explain the findings by stating that home and neighborhood environments must be more determinative of student outcomes than whatever happened in classrooms. All of this had the further effect of redirecting policy and media discussion away from race to class — thereby in turn eroding, however unintentionally, not only support for enrichment but also for desegregation. Additionally, the report, again inadvertently, called into question the conviction, so essential for advocates of desegregation and early enrichment alike, that children's brains were malleable and that changing their environments improved their IQs.

Chapter 2 turns to psychologists' involvement, in the course of the 1960s and 1970s, in diagnosing a condition in young children first labeled "hyperkinesis" and subsequently "minimal brain dysfunction" (MBD). Although asserted to be a physiological matter, one best treated with stimulant drugs like methylphenidate (Ritalin), the diagnosis was based on an observation of behavior and not on a clear medical symptom. Quite soon, and for a complicated conjunction of reasons, the modal individual for whom Ritalin became considered the most appropriate treatment was a white and middle-class child. As desegregation was often followed by the new phenomenon of tracking within schools, and as more African American children were labeled as suffering from "mild mental retardation" or placed in "slower" tracks, it became apparent that with the construction of a contrasting diagnosis of MBD a new disease entity had been created to address the cognitive challenges sometimes faced by privileged children of the predominantly white suburbs. Yet in the general upheaval of debate over race and intelligence convulsing the nation at the turn from the 1960s to the 1970s, well-meaning school psychologists urged that also minority children deserved the diagnosis of MBD and the concomitant preferred treatment with Ritalin. Or they contended that pills were necessary supplements to what might otherwise be ineffectual compensatory programming; some suggested that poverty itself affected young brains and caused MBD. Others, however, worried that an epidemiological finding that MBD might be prevalent among underprivileged African Americans could give fodder to racist positions on genetics and IQ. Simultaneously, a growing number of radical commentators, both African American and anti-racist white, came fiercely to protest what they perceived to be a disturbing tendency to overprescribe pills to poor children of color.

Chapter 3 concerns another area of argument within a nascent field of research that at first glance appeared to have little to do with race, class, or children's intelligence but soon enough became imbricated with all of these: a neuropsychological finding that humans had a bicameral brain. Split-brain theorizing became the lingua franca of the 1970s and 1980s, with the left hemisphere considered the seat of rationality and language while the right hemisphere housed intuition and creativity. Here again, as before with MBD, expert and popular writing on cerebral asymmetry came to be directed to society's privileged, who were encouraged to expand their right-brain potential with yoga, transcendental meditation, and biofeedback. At the same time, a substantial part of debates among neuropsychologists and related medical, social-scientific, and educational professionals revolved around the implications of such a revaluing of right-hemispheric skills specifically for African American, Latino, and Native American children. For the first time in the apparently interminable conflicts over race and intelligence, and rather than challenging biological determinists, a remarkable array of experts began to affirm the existence of racial differences in intelligence while taking up a critique that "right-brained" (and often poor and minority) children were trapped in "left-brained" schools. Declaring IQ to be an inaccurate measure, psychologist Alan S. Kaufman in 1979 developed an influential alternative assessment scale specifically to expand what counted as intelligence and to include a range of creative, nonverbal, spatial, and emotional capacities — only to find that gaps in test scores between white and nonwhite children narrowed accordingly.

The fourth chapter pursues an area of psychological investigation less inflected by countercultural impulses and more in tune with yearnings for life success felt across the ideological spectrum: the idea that self-control and ability to defer gratification was an essential skill infinitely more important than IQ. Already in the late 1950s, personality psychologist Walter Mischel had conducted experiments to assess children's ability to postpone the enjoyment of rewards; by the late 1960s, he had begun to test preschoolers on their ability to resist the temptation of one marshmallow available immediately in the expectation that if they waited they could have two marshmallows subsequently. Mischel saw his experiment as having a powerful predictive value, even as others — including Daniel Patrick Moynihan in his 1965 report on the African American family — began to racialize Mischel's findings in a way that pathologized black youth. Yet self-discipline was a far older ideal and had long been strongly race- and class-coded — even while earlier educational and psychological commentators had quite divergent understandings of its value. These earlier commentators worried, for instance, either that too much self-control resulted in unhealthful neuroses or that there were powerful class-determined (not racial-biological) reasons for individuals with truncated life prospects to opt for immediate (rather than postponed) gratifications. By the 1970s, however, psychologists (like Richard Herrnstein, future coauthor of *The Bell Curve*) announced that there were correlations between low impulse control, low IQ, and criminal behavior. By the mid-1990s, after the publication of *The Bell Curve*, Harvard-educated psychologist Daniel Goleman, in his best seller *Emotional Intelligence*, argued that self-control — not IQ — was most likely to predict future success. Emotional intelligence was to become a critically important alternative both to the metric of IQ and to the policy recommendations (and biological determinism) advanced in *The Bell Curve.*

Chapter 5 returns to efforts to justify early enrichment programming for underprivileged children that had exercised progressives since Head Start had been founded in 1965. After the appearance of *The Bell Curve* and its (however qualified) reintroduction of the science of biological inheritance, its statistical evidence of a persistent gap in IQ and achievement between African Americans and whites, and its overt critique of federal investments in compensatory programs, efforts to neutralize the book's impact focused on findings from neuroscience. In 2000 there came *From Neurons to Neighborhoods: The Science of Early Childhood Development*. This was an important and pathbreaking document that clarified how nature and nurture were in fact inseparable and — in contrast to *The Bell Curve* — that marshaled overwhelming neuroscientific evidence that the brain of a young child was plastic and amenable to the influences of environment; that inadequate nutrition, environmental toxins, and the stresses of poverty itself could harm a young brain; and that early intervention programming represented an ethical and moral imperative. *From Neurons to Neighborhoods* deemphasized race. The report chose instead to underscore the centrality of class, thereby avoiding some of the very problems the *Brown* decision had unintentionally produced. The report also sought strategically to foreground the view that the neurosciences were now critical to any policy discussions of early childhood.

Yet if *Brown* had shifted the conversation from material inequities to psychic intangibles, *From Neurons to Neighborhoods* shifted the conversation once again. The damages done by poverty were certainly real, the argument now became, but it was preeminently the neurosciences that would prove that they were. Sociological problems could be documented at the somatic level.

Yet although by the middle of the first decade of the twenty-first century, liberals may have thought that they, with their adroit deployment of neuroscience arguments, had won the day, the very same years saw the rise of a novel right-wing proposition: that it was *gifted children* who deserved more federal funding of enrichment programming instead of the poor (whether or not they were of color), on whom such monies were wasted anyway. At long last adopting for its own purposes the previously liberal preoccupation with the plasticity of young brains, the new right-wing trend contended that it was children with the highest IQs who required increased investments in their nurture. Meanwhile, on the other side of the ideological divide, proponents of early preschool provision for the impoverished came to worry that their own invocation of neuroscientific data to buttress claims that poverty caused demonstrable harm to young brains might inadvertently stigmatize further — as *Brown*'s emphasis on psychological damage had done sixty years earlier — the vulnerable children this evidence had initially been designed to help. The impasses in the liberal approach proved to be multiple and intractable, as right-wing views — amid a larger onslaught against public schooling — were once again in the ascendant.[23]

IT HAS BEEN ARGUED — and compellingly so — that the *Brown* decision had been motivated less by a longing to be ethically just and more by a desire to be politically expedient. The decision in *Brown* came at a critical moment in postwar U.S. history when the interests of middle- and upper-class whites and African Americans temporarily intersected. There were powerful Cold War imperatives to present the United States to the world as a bastion of democracy that stood in stark contrast to life behind a communist Iron Curtain.[24] Related was the fact that racial segregation came increasingly to be seen as an albatross for U.S. foreign policy, especially in dealings with newly independent nations on the decolonizing continents of Africa and Asia.[25] Additionally, pressures mounted from an African American community energized by its many veterans who returned from the fight against German fascism only to find themselves in a fight against white supremacy at home.[26] And finally, the economic incentives associated with racial segregation had steadily been in decline during the postwar period, as Jim Crow policies came to be perceived by white elites as an impediment to corporate profits.[27] The perspectives of legal scholar Derrick Bell remain the clearest; Bell wrote that the urge to do the right thing was not enough to push the nation to abandon de jure segregation. "As with abolition," Bell observed, "the number who would act on morality alone was insufficient to bring about the desired racial reform." Bell added that an old American principle remained operative in *Brown*: "The interests of blacks in achieving racial equality will be accommodated only when it converges with the interests of whites." Bell soberly concluded that "the fourteenth amendment, standing alone," was not enough to "authorize a judicial remedy providing effective racial equality for blacks where the remedy sought threatens the superior societal status of middle and upper class whites."[28] Or as Bell would also put it in "a sardonic formula for what I had come to understand as the basic social physics of racial progress and retrenchment": "Justice for blacks vs. racism = racism. Racism vs. obvious perceptions of white self-interest = justice for blacks."[29]

Meanwhile, and no matter how equivocal its motivations may have been, or how unfulfilled its hopes and promises turned out to be, *Brown* was to have yet further unanticipated consequences. Among other results, there would be *Brown*'s troubling — if not outright catastrophic — spur to white racial consciousness. Close to two generations after *Brown*, civil rights activist and legal scholar Michelle Alexander in *The New Jim Crow*, a study of the mass incarceration of African American men, voiced her deep sorrow that in a post-*Brown* United States, "priority should have been given to figuring out some way for poor and working-class whites to feel as though they had a stake — some tangible interest — in the nascent integrated racial order."[30] Instead, as legal scholar Michael Klarman provocatively wrote, *Brown* "crystallized southern resistance to racial change," even as it "propelled politics in virtually every southern state several notches to the right on racial issues."[31] And as legal scholar Lani Guinier observed, particularly for working-class whites, "compulsory association with blacks brought no added value and endangered the sense of autonomy and community they did have. *Brown*'s racial liberalism did not offer poor whites even an elementary framework for understanding what they might gain as a result of integration."[32]

As the chapters that follow will show, throughout the postwar years, and long before the turn to the neurosciences, psychology as a discipline and as a popular phenomenon provided an extraordinarily significant way both to grapple with some of the most intense, anguishing, and unresolved issues in American life and to distract attention away from those same issues. One of the great dilemmas of politics in our present is that middle-class and poor whites are also suffering, as they too are increasingly economically and existentially destabilized. It has become dismayingly fashionable to emphasize this point. But decades of lies have led many whites to misdirect their anger at the nonwhite poor.[33] So much public commentary and so many policy discussions about race, class, education, and intelligence in the United States since the *Brown* decision have served the double purpose of, on the one hand, soothing white self-esteem and, on the other, promoting excuses to decimate public education and public expenditure more generally. It is hoped that a history of popular debates over race and intelligence and the policy consequences of psychological and neuroscientific theorizing will make clear how woefully wrongheaded the misdirection of that anger is.

ONE

Plasticity of Intelligence, Education Reform, and the Disadvantaged Child

O N FEBRUARY 19, 1969, in his first major address to Congress, President Richard M. Nixon sought to allay fears that he intended to eradicate anti-poverty programs established by his predecessor, Lyndon B. Johnson. Yet Nixon also voiced profound disappointment at what he said were the "shortcomings" of federal efforts to end "the blight of poverty." In particular Nixon took aim at Project Head Start, the federal program designed to assist impoverished preschool children, noting how "the results of a major national evaluation of the program . . . confirm what many have feared: the long term effect of Head Start appears to be extremely weak." Nixon strikingly asserted that while "intelligence is not fixed at birth" and "is largely formed by the environmental influences of the early formative years," it still remained true "that much that has been tried has not worked" and that "these past years of increasing Federal involvement have begun to make clear how vast is the range of what we do not know."[1] In this way, Nixon acknowledged the prospect that a child's cognitive capacities *could* be improved through early educational interventions. But he stressed that for all practical purposes, such change was highly difficult, if not virtually impossible, to achieve.

Less than two months later, on April 9, 1969, Nixon elaborated on these pessimistic perspectives. He declared that "preliminary evaluations" indicated that "Head Start must begin earlier in life, and last longer, to achieve lasting benefits." While this comment would seem to indicate the need for *more* funding rather than less, and while he reiterated that "Head Start appear[s] to come too late for many of those children who most need help" and that "the process of learning how to learn begins very, very early in the life of the infant child," Nixon went on to suggest that all compensatory efforts to advance a child's cognitive abilities made little difference in that child's life. As Nixon put it, the "children of the poor" were "handicapped as surely as a child crippled by polio is handicapped, and he bears the burden of that handicap through all his life," and therefore Head Start children were unlikely to benefit in any meaningful way from the program. It was in this context too that Nixon cited to a national audience a psychological concept to explain how impoverished children often became permanently "crippled" and "handicapped" in their capacity to learn. Nixon posited that "an impoverished environment can develop a 'learned helplessness' in children." He added that "when there is little stimulus for the mind, and especially when there is little interaction between parent and child, the child suffers lasting disabilities, particularly with respect to the development of a sense of control of his environment."[2] When children lagged in cognitive development and academic achievement, the root causes of these problems lay in realities that schools could not alter; therefore the impact of improved schooling on these children's intelligence would always remain comparatively minor.

These expressions of concern with respect to the damages done by socioeconomic inequality, if taken out of context, could make Nixon sound like a Marxist. Instead, however, Nixon's references to psychological theory — to "learned helplessness" and its contrast with "a sense of control"— provided an intellectual framework for his administration's subsequent policy efforts to defund educational enrichment programs. Among these were projects like Head Start but also compensatory programming at the elementary and secondary levels. The argument became that these efforts were both needlessly costly and essentially ineffective.[3]

Nixon's bald assertion that "learned helplessness" in poor children demonstrated the futility and wastefulness of compensatory educational efforts brought psychological research directly into political and policy discussions of race, class, and child development. And even as he avoided direct mention of race in this context, Nixon nonetheless allowed his references to Head Start to evoke images of poor children of color.[4] In this way, Nixon's nod to psychological study actually represented a blunt attempt to *turn back* two decades of psychological research that had argued that it was the quality of schooling that mattered most for the cognitive development of young children and that children of whatever socioeconomic status or racial background stood to thrive in conditions where they were consistently encouraged to be motivated to learn. It may have been a consensus in child development research during the earlier decades of the twentieth century that human intelligence was fixed at birth, but that consensus had been strongly challenged between the late 1940s and the early 1960s.[5]

By the beginning of the 1960s, confidence was running especially high that educational interventions in the preschool years could accelerate the cognitive abilities of disadvantaged children. In 1961, for instance, University of Illinois educational psychologist Joseph McVicker Hunt forcefully wrote that it was entirely "feasible to discover ways to govern the encounters that children have with their environments, especially during the early years of their development, to achieve a substantially faster rate of intellectual development and a substantially higher adult level of intellectual capacity."[6] Hunt—who became one of the major rediscoverers during the 1960s of Maria Montessori's theories about how children learn—specifically meant his research to be applied to poor and nonwhite children. In 1962, at a conference on the "pre-school enrichment of socially disadvantaged children" and framed by a discussion of Montessori's "largely forgotten work," Hunt elaborated on a set of positions that he argued should orient a new generation of psychological research into a child's cognitive development: that human intelligence was remarkably malleable; that social environment therefore had a profound impact on a child's capacities for cognitive and emotional growth; and that early intervention efforts could absolutely spur the development of intelligence and thus could well serve as "an antidote for cultural deprivation."[7] In 1963, developmental psychologist David P. Ausubel declared, "The possibility of arresting and reversing the course of intellectual retardation in the culturally deprived pupil depends largely on providing him with an optimal learning environment as early as possible in the course of his educational career."[8] And in 1964, educational psychologist Benjamin S. Bloom was even more precise, noting that "in terms of intelligence measured at age 17, about 50% of the development takes place between conception and age 4, about 30% between ages 4 and 8, and about 20% between ages 8 and 17."[9] Bloom, together with psychologists Allison Davis and Robert Hess, contended—in a report based on a 1964 gathering sponsored by the U.S. Office of Education of more than two dozen social scientists—that because a child "'learns to learn' very early," it was important that "instruction is individual and is timed in relation to experiences, actions, and questions of the child" and, moreover, that it would be desirable that the child "comes to view the world as something he can master through a relatively enjoyable type of activity, a sort of game, which is learning."[10] This model of early childhood learning that emphasized flexibility and adaptability was to have profound consequences on policy decisions to establish early educational opportunities for poor and minority children.

Historians of social policy in the postwar era have repeatedly observed that the scholarly arguments advanced by Hunt, Ausubel, Bloom, Davis, Hess, and other influential psychologists served as "the scientific and intellectual basis" for federal anti-poverty preschool initiatives after the passage of the Civil Rights Act in 1964.[11] An advocate of these programs was matter-of-factly to comment only a few years later, "By the middle of the sixties, no thinking person could ignore the importance of the first few years of life for subsequent developmental competence."[12] It was this liberal (and boldly optimistic) social scientific consensus about the remarkable plasticity of human intelligence and the reversibility of the deleterious effects of "cultural deprivation" that Nixon now sought only a handful of years later at the end of the decade — citing new (if vague and unspecified) research on "learned helplessness"— to destabilize and disrupt. Thus in a strategically complicated double maneuver, Nixon took up strands of psychological research only to recombine them in new ways, simultaneously citing the behavioral science of "learned helplessness" *and* reaffirming the plasticity of human intelligence in order to challenge, or at least to complicate, the idea — which he presented as an outmoded position — that the intelligence of disadvantaged children would be benefited through compensatory educational strategies.

In the twenty-first century, "learned helplessness" has largely been a tainted terminology. Its coinage is chiefly attributed to behavioral psychologist Martin E. P. Seligman, later the founder of the immensely successful phenomenon of positive psychology. But "learned helplessness" has most recently become closely associated with the psychological underpinnings of a covert CIA "enhanced interrogation" program that engaged in the torture of terror suspects — as well as with the controversies surrounding Seligman's possible complicity in helping to advise the developers of this program.[13] Yet when Nixon spoke of "learned helplessness," the phrase had no popular or political resonance whatsoever. In the spring of 1969, it remained an empty concept. Articles on the subject were just then beginning to be circulated, but only in the most specialized academic journals (such as the *Journal of Comparative and Physiological Psychology* and the *Journal of Experimental Psychology*).[14] Indeed, the first general interest essay on the topic — by

FIGURE 2. President Richard M. Nixon with Daniel Patrick Moynihan, executive secretary of the president's Urban Affairs Council, on the Washington Mall on September 8, 1970. Earlier that year, Moynihan had written to the *Journal of Social Issues* to dispel a report that he had given Arthur Jensen's writing on race and intelligence "to the Nixon Cabinet as 'must reading.'" Moynihan averred, "Nothing of the kind happened," though he added, "I know what Jensen is going through. I got the same treatment for almost exactly the opposite hypothesis!" Reprinted by permission of AP Images.

Seligman — appeared in *Psychology Today* only two months *after* Nixon's speech.[15] Nixon's reference to the behavioral science of learned helplessness remains striking because there was no science. The only empirical evidence that *did* exist on learned helplessness had nothing whatsoever to do with the behavior of human beings — let alone of disadvantaged children. Studies in learned helplessness by the spring of 1969 were entirely restricted to laboratory experiments conducted with albino rats and mongrel dogs.[16] It was Nixon's idea (or at least the idea of his speechwriters, including the president's adviser on urban affairs, Daniel Patrick Moynihan) to forge a link between the study of learned helplessness and the prospects of the disadvantaged child (see figure 2).

Yet as the several references from educational psychologists already cited suggests, Nixon's selective turn to the psychological in 1969 was not the beginning of the story of how psychological research in the decade informed national discussions of race, class, and the cognitive development of disadvantaged children. This conversation had by this time been ongoing, as it sought answers to a variety of interrelated questions: If disadvantaged children failed to learn, where did their difficulties with academic achievement originate? Was it the result of their class background — or was it due to racial discrimination? Could it be traced to the lesser quality of their school facilities? Or did their difficulties with learning stem principally from problems in their family life and home environment? If home and family were the cause, could school desegregation make a positive difference in the child's academic achievements and motivational skills? And if the problem was due to race and class, would compensatory preschool programs result in lasting cognitive gains — or would the gains of disadvantaged children "fade out" or "level off" after they had left these programs (as critics of compensatory programs came increasingly to assert)? And what finally of the role played by the child's teachers? Did teachers' low expectations for the intellectual potential of disadvantaged pupils condemn these students to failure? Or conversely, did these students in fact condemn themselves to failure by having too little faith in their own abilities? These final two queries — in tandem and separately — came in particular to generate remarkable national discussion of public policy and considerable scholarly debate in the course of the 1960s.

The Pygmalion Effect

Why did disadvantaged children so often fail to learn? One commonsense answer was that their teachers expected far too little of them and that these students' failure to achieve amounted to a self-fulfilling dynamic caused by teachers' attitudes. In the spring of 1969, just when Nixon was speaking of "learned helplessness," there was in wide circulation discussion of another psychological phenomenon officially known as an "interpersonal expectancy effect," which came far more routinely to be called the "Pygmalion effect." The Pygmalion effect had been named by social psychologist Robert Rosenthal and educator Lenore Jacobson after George Bernard Shaw's play *Pygmalion* and its tale of Eliza Doolittle, a poor and uncouth flower girl, whom Henry Higgins, a pompous professor of phonetics, had wagered he could transform into a genteel, aristocratic lady. After her transformation has been achieved, Eliza informs the kindly Colonel Pickering (who had arranged the wager with Higgins), "You see, really and truly, apart from the things anyone can pick up (the dressing and the proper way of speaking, and so on), the difference between a lady and a flower girl is not how she behaves, but how she's treated. I shall always be a flower girl to Professor Higgins, because he always treats me as a flower girl, and always will; but I know I can be a lady to you, because you always treat me as a lady, and always will."[17] In 1964, this speech of Eliza's had just been included — almost verbatim — in the highly acclaimed film adaptation of Shaw's play, *My Fair Lady*, even as it effectively summarized what Rosenthal and Jacobson believed about the central role that a teacher's expectations of a student had on that student's capacity to learn.

There were, Rosenthal and Jacobson proposed, powerful "Pygmalion effects" in the American classroom; schoolteachers who *expected* their students to make cognitive gains saw these students make those very same gains. A student's intelligence could be improved when teachers believed that the student was about to experience a spurt in cognitive growth. Meanwhile, and conversely, if teachers identified a more limited potential in some of their students, those students underperformed.

Rosenthal's research initially had nothing to do with the impact of teachers' expectations on the intellectual abilities of children — disadvantaged or otherwise. What had interested Rosenthal since the late 1950s were the self-fulfillment principles actively at work in the (only seemingly) controlled setting of the psychological laboratory. In the early 1960s Rosenthal had instructed students in a class he was teaching on experimental psychology (at the University of North Dakota) to replicate experimental findings with albino rats. He informed his students that "studies have shown that continuous inbreeding of rats that do badly on a maze leads to successive generations of rats that do considerably worse than 'normal' rats," while "generations of maze-bright rats do much better than generations of maze-dull rats." Despite the fact that this was utter hokum and that not one of the rats had demonstrated any prior differential talents in ability to navigate a maze, the students nonetheless managed to produce data that rats labeled "maze-bright" had learned the maze significantly faster than had rats labeled "maze-dull."[18] Rosenthal wrote, "The basic paradigm for the study of this phenomenon has been to create two or more groups of [experimenters] with different hypotheses, or expectations about the data they would obtain from their [subjects]."[19] Furthermore, it was Rosenthal's strong speculation that the "maze-bright" rats had received subtly preferential treatment and handling from the assistants.[20] The students viewed the "maze-bright" rats as being "more pleasant" and "more likeable" than the "dull" rats and so treated these "maze-bright" rats with greater warmth, gentleness, and enthusiasm than they had treated the "dull" rats.[21] The rats had, measurably, done better or worse as a result of the experimenters' expectations.

Rosenthal's experimental research with laboratory animals has merited no mention in histories of psychology's engagement with race in the postwar era. Instead Rosenthal's research with rats has been interpreted as part of a larger and very important cluster of psychological studies in the early 1960s that sought to bring into view the ways in which psychology experiments can suffer from "experimenter bias." Indeed, it represents one of the most interesting examples from the era of critical, self-reflexive efforts within various branches of psychology that sought to grapple with problems inherent in the profession's experimental methods.[22] Yet Rosenthal's rat studies soon became most thoroughly enmeshed in discussions of race precisely because he himself began to observe how his animal experiments should be generalized to grasp how "disadvantaged" children of color learned how to learn — or, far too frequently, failed to do so. Rosenthal recognized almost immediately that his research with animals had possible applications for schoolchildren. Already in the same 1963 report on "experimenter bias" in research with rats, Rosenthal asked, "When the master teacher tells his apprentice that a pupil appears to be a slow learner, is this prophecy then self-fulfilled?"[23] It was this question that Rosenthal soon set himself to answer, with a particular interest in how teachers' expectations, whether positive or negative, often came to shape student achievement — including, crucially, the measurability of that achievement in the form of IQ.

What happened was that Lenore Jacobson, an elementary school principal in the South San Francisco Unified School District, came across Rosenthal's animal research on the effect of expectations and wondered if the experiment could be tested with children. At that time, Jacobson was completing a Berkeley dissertation on the ways in which public schools all too often impeded the scholastic potential of Mexican American children. In 1964 Jacobson invited Rosenthal (by this point at Harvard) to her elementary school to carry out a study to explore why it was that lower-class Latino children — her school's most prevalent minority group — tended "to overpopulate the slow track and underpopulate the fast track."[24] The experiment they designed closely modeled the laboratory studies Rosenthal had previously conducted with rats. Rosenthal and Jacobson informed the school's teachers in May 1964 that a "Test of Inflected Acquisition" would be administered to students by Harvard investigators and that this test could predict whether a student was about to "bloom" intellectually. (The test had no such predictive capacity.) In September 1964 at the first staff meeting for the new school year, Rosenthal and Jacobson notified the school's teachers that approximately one out of every five of their students had done exceptionally well on the "test for intellectual blooming" and so would be expected in that coming academic year to demonstrate "unusual intellectual gains" (although these children had been randomly chosen).[25] At the end of the school year, Rosenthal and Jacobson reviewed the data based both on actual IQ tests the students had taken as well as on self-reports from teachers. First, they found that students who had been tagged as about to "bloom" in their intelligence had done well with their teachers; when asked to describe their students, teachers singled out "the children from whom intellectual growth was expected" (whether white or Latino) as "having a better chance of being successful in later life and as being happier, more curious and more interesting than the other children."[26] Second, they found that these same favored students — especially the first-graders — had made significant gains on their IQ tests in relation to the control group of non-favored students. And third, Rosenthal and Jacobson found that the Latino students had experienced "greater benefits of favorable teacher expectations than had the white children." However, they acknowledged that although Latino children "had shown greater benefits of favorable teacher expectations" than had the non-Latino students, these differences "did not reach statistical significance."[27] The empirical evidence that teachers' often subtle and unspoken behavior was having a demonstrable impact on minority students' gains in IQs, in short, remained tantalizing but incompletely conclusive. Nonetheless, Rosenthal and Jacobson continued to be convinced that race as well as class mattered greatly when it came to the differential treatment meted out by many (white) schoolteachers. In other words, they believed in the detrimental consequences of unconscious racism — what would in present-day language more likely be termed "implicit bias."[28]

The culmination of Rosenthal and Jacobson's collective efforts appeared in 1968 with the provocative title *Pygmalion in the Classroom: Teacher Expectation and Pupils' Intellectual Development*. The *Elementary School Journal* encapsulated the book's core thesis with this pithy phrase: "What Teachers Believe — What Children Achieve."[29] *Time* magazine reported that the Rosenthal and Jacobson book proved that "many children fail to learn simply because their teachers do not expect them to."[30] Rosenthal further publicized his arguments during an appearance on *The Today Show*, where he was interviewed by Barbara Walters. According to Stanford educational psychologist Nathaniel L. Gage, *Pygmalion in the Classroom* went on to receive "more attention in the mass media than any other product of the behavioral sciences in the 1960's"—an exceptional accomplishment given that the decade saw quite a number of influential studies in the behavioral sciences, including Albert Bandura's experiments in social modeling, Stanley Milgram's experiments in obedience to authority, and John Darley and Bibb Latané's experiments on bystander apathy, among others.[31] As for the book's cultural and political impact, social psychologist Roger Brown later recounted that many liberals in the educational and psychological communities at the time began to speculate whether "the negative expectancies of teachers might be the whole explanation of the low academic achievement of various ethnic minorities."[32]

Pygmalion in the Classroom's perspective was hardly unprecedented. It was not new news to suggest that teacher expectations shaped students' academic performance, especially in the context of a historical moment when many psychologists and educators routinely argued that a classroom's environment had the potential to devastate especially the disadvantaged child. In *How Children Fail*, his best-selling diary about teaching the fifth grade in Cambridge, Massachusetts, educator John Holt sketched the often subtle and unintended but nonetheless terrible ways that teachers perpetuated feelings of defeat and shame in their students. Anticipating Rosenthal and Jacobson and striving to explain how underachieving students could, over time, have developed habits and patterns "of weakness, of incompetence, of impotence," Holt reasoned that the cause might lie in teachers' attitudes. Imagining how the teachers' approval or disapproval might well undermine students' capacity to learn ("For, after all, if *they* (meaning we) know that you can't do anything, *they* won't expect you to do anything, and *they* won't blame you or punish you for not being able to do what you have been told to do"), Holt concluded, "Perhaps, in using the giving or withholding of approval as a way of making children do what we want, we are helping make these deliberate failers."[33] Early 1960s commentators often offered comparable arguments to provide a context for the specific academic deficits then found to be more common among African American children. Journalist Charles Silberman in his best-selling *Crisis in Black and White* of 1964 observed that regardless of whether the "prejudice" of a white teacher was "conscious or unconscious, it presents a terrible block to learning. Children, no less than adults, resent being patronized. Equally important, young children tend to fulfill the expectations their elders hold out for them. Hence the teacher who assumes that her children cannot learn very much will discover that she has a class of children who indeed are unable to learn—a prime example of what sociologists call a 'self-fulfilling prophecy.'"[34] Pediatrician Benjamin M. Spock wrote in the same year that when "a teacher believes that a certain student—of whatever color—is stupid, even though he really has a satisfactory aptitude," that student may well respond by "seeming to ignore the teacher and the teaching material."[35] Psychologist Kenneth Clark— coauthor of the "doll experiments" that had provided crucial support for the *Brown* decision—stated at a 1965 conference on "environmental deprivation and enrichment" that it was possible that "social deprivation theories" were merely "offering an acceptable alibi for pervasive educational default, a default in which the public school system permits and perpetuates educational inefficiency when dealing with powerless children or children from powerless roots in the society." Of disadvantaged youth, Clark further inquired, "To what extent are they not being taught because those who are in charge of teaching them do not believe that they can learn, do not expect that they can learn, and do not relate to them in ways that are conducive to their learning?"[36]

There was as well a widely circulated (though possibly apocryphal) anecdote about an unpleasant white New York City public schoolteacher who routinely spoke to the African American students in her classroom "in a belittling manner," only then to state out of their earshot "that 'heredity is what really counts,' and since they don't have a high culture in Africa and have not yet built one in New York, they are intellectually inferior from birth."[37] In 1966 educational psychologist Arthur Pearl succinctly summarized the problem: "At the heart of academic failure is poor teaching."[38] In addition, the 1967 U.S. Commission on Civil Rights study *Racial Isolation in the Public Schools* argued how it was no secret that teachers needed "to improve both their expectations of the students and their ability to teach disadvantaged children."[39] And there was the appalling and telling story that St. Louis school district superintendent Samuel Shepard Jr. related in testimony to a Senate subcommittee hearing on civil rights:

> Here was a teacher who had copied the IQ numbers down the line from a list in the principal's office. . . . Throughout the semester if the teacher called on Mary, let us say, with an IQ of 119, she followed this pattern:

if Mary didn't respond quickly, "Well, now, come, Mary. You know you can do this. You know how we did this yesterday," or bring up an analogous situation. However, when she called on poor John with his 74 IQ, if he mumbled something, fairly audible, why, this was wonderful; pat him on the back and [say], "Be sure and be here tomorrow. You can wash the windows and help move the piano and water the flowers, and the erasers must be washed," and so forth. This is the kind of encouragement that he got with a 74. This is teaching by IQ. She was a little horrified at the end of the semester when she turned in her grades. She looked under the glass and saw that the columns she had copied for IQs were locker numbers.[40]

So when Rosenthal began pointedly to observe that if an elementary teacher "is to teach a 'slow group,' or children of darker skin color, or children whose mothers are 'on relief,' she will have different expectations for her pupils' performance than if she is to teach a 'fast group' of children of an upper middle class community," he was expressing sentiments fully in step with the convictions of his moment.[41] Rosenthal and Jacobson's argument was broadly held among educators and psychologists.

Also, empirical support for Rosenthal and Jacobson's position that "self-fulfilling prophecies in the classroom" would redound to the benefit of students' achievement levels when teachers communicated that they believed in their students' intellectual capabilities preexisted their Pygmalion study. For instance, there was the Demonstration Guidance Project, first established in 1956 in a predominantly African American and Latino junior high school in Harlem (under the guidance of Kenneth Clark); it expanded by the early 1960s to dozens of schools, whereupon it was renamed the Higher Horizons Program for Underprivileged Children. Higher Horizons coordinator Daniel Schreiber said in 1961 of his project's philosophy of education, "Our basic approach is to create in the mind of a child an image of his potential, [and to] fortify this image by parent, teacher, and community attitudes."[42] The skill level of Higher Horizon students was judged that year to have markedly improved: "An evaluation of the program found that 147 of 250 students who had begun the project in seventh grade gained on the average 4.3 years in reading achievement after 2.6 years of the program at the junior high school."[43] Another example was the Banneker Project in St. Louis, supervised by Samuel Shepard Jr. and begun during the 1957–58 school year, specifically targeted at elementary school children from impoverished backgrounds. Shepard's philosophy was straightforward: "I have an abiding faith that our children have the ability to learn."[44] His initial results too were impressive; the *Southern Education Report* in 1965 announced that "the average intelligence quotient of the children in the Banneker District was raised 11.5 points and children were brought up to the national norm in language, writing and arithmetic."[45] In *Dark Ghetto*, Kenneth Clark wrote stirringly of the Banneker Project's accomplishments: "In spite of the fact that there had been no drastic change in curriculum, instructional technique, or the basic 'underprivileged' social situation, improvements were definitely evident. What had changed was the attitude and perspective of teachers which influenced the way in which students were taught and learned."[46] At least for a time into the mid-1960s, a conviction about the plasticity of children's intelligence and the capacity to reverse the effects of poverty and cultural deprivation seemed to be supported by facts.

Locus of Control and the Coleman Report

Why did disadvantaged children so often fail to learn? A second common-sense answer by the mid-1960s was that students expected far too little of themselves and that these students' difficulty with achievement stemmed from a problem of self-concept. This was the view that President Nixon advocated in 1969. Nixon acknowledged the plasticity of child intelligence but uttered not a word about a possible Pygmalion effect in the classroom. Instead he spoke of the child who lacked "a sense of control of his environment" and who was therefore "handicapped" academically as a result. Nixon's view of schools had nothing to do with the notion that schoolteachers often enacted (what Arthur Pearl labeled) "ceremonies of humiliation" on their students.[47] Rather, it had almost everything to do with an alternate view that focused more exclusively on what students *believed about themselves.*

This perspective could be traced back to the work of clinical psychologist Julian Rotter. In 1954 Rotter first proposed that the outcome of past events inevitably came to have a more general set of consequences for what a person came to expect about the outcome of future events. Rotter wrote, "What happens to the child's expectancy of getting an ice cream cone after he has just been refused a candy bar?" To which he commented, "It can be demonstrated that the occurrence (or non-occurrence) of a given reinforcement produces changes in expectancy for the occurrence of other reinforcements."[48] He divided all people into two general categories: those who possessed an "external locus of control" (defined by Rotter as the tendency to interpret personal successes and failures as the result of luck or chance) and those who possessed an "internal locus of control" (defined as the tendency of persons to believe themselves to be masters of their own fate). Rotter hypothesized that whether a person was oriented "externally" or "internally" was "of major significance in understanding the nature of learning processes in different kinds of learning situations and also that consistent individual differences exist among individuals in the degree to which they are likely to attribute personal control to reward in the same situation."[49] Rotter soon expanded his conception of expectancy still further to the issues of race, class, and learning.

In 1963, Rotter and fellow clinical psychologist Esther Battle reported on a study they had conducted with eighty sixth- and eighth-grade public school children — both African American and white, both middle class and poor — in the cities of Columbus and Dayton, Ohio. Rotter and Battle developed a projective test to measure these children's external versus internal loci of control. The test used cartoon pictures of figures in a variety of situations; the child was then asked to answer a question about each picture, such as "Why is she always hurting herself?" or "Why does her mother always 'holler' at her?" What Rotter and Battle's evidence determined was that the children who were (in their terminology) "internals" were most likely middle-class and white; children who were "externals" were more likely poor and black — although middle-class African American children were more "internal" than were poor black children. Significantly, Rotter and Battle found, on the one hand, that "middle-class children, in general, were significantly more internal than lower-class children"— regardless of race — suggesting that class (not race) determined "locus of control," even as, on the other hand, they found that poorer blacks with high IQs were "more external" than middle-class whites with lower IQs, suggesting that race (not intelligence or class) determined "external attitudes" (possibly, they suggested, "as a defense reaction to perceived reduced choices for cultural or material rewards"). "These results suggest that one important antecedent of a generalized expectancy that one can control his own destiny is the perception of opportunity to obtain the material rewards offered in a culture," reiterated Rotter and Battle.[50] A correlation between poor and minority children's inability to have any faith that they were the arbiters of their own destiny and these children's lack both of motivation and achievement began from this moment to inform much further psychological inquiry.

Throughout the first half of the 1960s, clinical psychological research found that the "external" orientation said to be more common among racial minorities corresponded with feelings of personal powerlessness, while "internal" orientation more frequently noted among whites corresponded with a stronger sense of personal mastery and control. (It remains unclear how much experimenter expectancy may have itself shaped these findings.) However, many of these studies examined young adults rather than children. For instance, a 1963 study of southern African American college students determined that those students who possessed a stronger sense of internal control had a stronger commitment to civil rights activism.[51] A second study conducted in 1965 surveyed more than a hundred African Americans in the Student Nonviolent Coordinating Committee and contrasted their responses with those of non-activist black students; its results "confirm the hypothesis, with Negro students who were known to be active in civil-rights demonstrations being significantly more internal than Negro students who had had no experience in protest movements."[52] Also in 1965, clinical psychologist Herbert Lefcourt, another early and important theorist of "locus of control," turned his attention to sixty white and sixty African American (and all mainly lower-class) inmates in two correctional institutions. Lefcourt rated these inmates' reactions to statements like "I have often found that what is going to happen will happen" and "As far as world affairs are concerned, most of us are the victims of forces we can neither understand nor control." Lefcourt found that African American inmates had a far greater pessimism — and a much higher measure of external control — than did the white inmates. He additionally posited that the "externally oriented Negro" likely had a tendency both to exert "a minimum of effort to achieve" and to demonstrate "a lack of interest in achievement-related pursuits" and that these tendencies could well mean "that Negroes' poorer performance on intelligence tests reflects a withdrawal from middle-class achievement goals."[53] Here too we see the endorsement of a (self-understood as liberal) perspective that intelligence was socially shaped and therefore alterable — even as, in Lefcourt's framework, what mattered most was the individual's sense of personal control over the environment (rather than the kind of "interpersonal expectancy" and attention to the attitudes of authority figures that had informed the Pygmalion framework).

Strikingly, Lefcourt cited the essays and novels of Richard Wright, Ralph Ellison, and James Baldwin when he argued, in another text published in 1965, that black men were "failure avoidant and behaviorally deviant largely because they have low expectancies for obtaining positive reinforcement for more socially desirable behavior." Here when Lefcourt reported on "a risk-taking experiment" conducted with black and white inmates who volunteered to participate in games both of chance and of skill, he found African American subjects "more highly motivated to avoid failure in skill situations and more motivated to achieve success in chance situations." This finding, Lefcourt posited, could be due to an American history of "segregation and discrimination practices" and a social universe in which black persons "in the schoolroom, market place, or job" find that their "work or effort will rarely lead to valued goals such as social recognition, financial comfort and security, protection from malevolent community forces, etc. Consequently, many Negroes have adopted less acceptable but relatively more risky methods for gaining certain goals."[54] In this way, Lefcourt sought to offer a sympathetic and context-sensitive reading of the African American personality.

Lefcourt's speculative claim that racial discrimination caused African Americans to be less motivated to achieve in realms of existence (like the classroom) where individual effort mattered greatly was to come to prominence in the context of the *Equality of Educational Opportunity* report in 1966. It had been the Civil Rights Act of 1964 — passed a decade after the *Brown* decision — that had required that a survey and a report be produced within two years that would analyze "the lack of equal educational opportunities for individuals by reason of race, color, religion, or national origin in public education institutions at all levels in the United States."[55] Based on questionnaires distributed in the fall of 1965 to four thousand public schools (although close to a third of the schools did not participate) and to more than 645,000 students in the third, sixth, ninth, and twelfth grades, the resulting seven-hundred-page report — which popularly came to be known as the Coleman Report (after its lead author, Johns Hopkins sociologist James S. Coleman) — was understood in its own day as having "turned understanding of a major area of social policy upside down as perhaps no comparable event in the history of social science."[56] Into the twenty-first century, the Coleman Report (whatever methodological limitations it may have had) remains recognized as "one of the most influential investigations ever conducted on education in the United States."[57]

In the same year, as the data for the Coleman Report was being collected, a pathbreaking bill was passed. President Johnson hailed the Elementary and Secondary Education Act (ESEA) of 1965 as a cornerstone of his administration's "War on Poverty." The dramatic goal of the ESEA was — as it was later summarized—"improving the academic achievement" of every "disadvantaged" schoolchild in the country. Indeed, more than half a century later, educators still call the ESEA "the most important piece of education legislation in U.S. history."[58] At the signing ceremony for the $1.3 billion education bill, Johnson stated that he envisioned that the ESEA would serve ultimately to "bridge the gap between helplessness and hope for more than five million educationally deprived children."[59] In the span of its first year, the ESEA more than doubled federal expenditures to public schools in the United States—with most of these funds directed to compensatory educational programs.[60] As for the Head Start Project initiated by the ESEA, Johnson announced in 1966 that the program—no matter the brevity of its existence—had already turned out to be "battle-tested and proved worthy."[61] The Coleman Report was supposed to buttress federal efforts that involved the pouring of monies into preschool enrichment programs as well as into the public schools more generally.

It was widely anticipated that the Coleman Report would determine not only that the desegregation of U.S. public schools as mandated by the *Brown* decision had continued to lag badly but also that segregated black schools still suffered from inferior physical and pedagogical resources—and that such differences accounted for the poorer academic achievements of black children. (Meanwhile, the possibility that some—even many—African American students had benefited greatly from having African American teachers and role models was not even considered.)[62] During the time that data for the *Equality of Educational Opportunity* report was still being collected, James Coleman expressed complete confidence that the data would confirm what he (and almost every other social scientist and educator) already took to be self-evident: a school's quality had direct bearing on the academic achievements of its students, and white schools were simply assumed to be of better quality. Coleman articulated this expectation explicitly in the fall of 1965 when he observed that "the study will show the differences in the quality and resources of schools that the average Negro child and the average white child are exposed to." Addressing his interviewer, he continued, "You know yourself that the difference is really going to be striking. And even though everyone knows there is a lot of difference between suburban and inner-city schools, once the statistics are there in black and white, they will have a lot more impact."[63] The stronger a school, the better its students achieved. And African American students were still by and large attending the nation's worst schools. The data would bear this out.

However, the Coleman Report's findings did not do what had been anticipated. The findings did show that desegregation lagged. African American
and white students did attend separate public schools, for the most part.
The report also showed that there existed significant gaps in the academic
performance of black and white children as measured by standardized test
scores in verbal ability, mathematical skill, and reading comprehension. Yet
it determined — and this was the complete surprise — that the available resources in black and white schools were more similar than not. White and
nonwhite students had approximately the same number of books in their
school libraries, and per student expenditures (for textbooks, audiovisual
equipment, and the like) was about the same for white and nonwhite children. Class size did not vary much, either. The report went on to propose —
and this is how it still principally remains remembered — that differences in
racial achievement could be traced not to differences in schools but rather
to differences *in homes*. "Studies of school achievement have consistently
shown that variations in family background account for far more variation
in school achievement than do variations in school characteristics," the report stated in its section on pupil achievement and motivation. The quality
of a school turned out *not* to be the key to its students' academic achievement after all. Instead the report controversially stated that "one implica-
tion stands out above all: That schools bring little influence to bear on a
child's achievement that is independent of his background and general social context; and that this very lack of an independent effect means that the
inequalities imposed on children by their home, neighborhood, and peer
environment are carried along to become the inequalities with which they
confront adult life at the end of school." The report continued, "For equality
of educational opportunity through the schools must imply a strong effect
of schools that is independent of the child's immediate social environment,
and that strong independent effect is not present in American schools."[64] In
short, the fact that the Coleman Report put forth a conclusion that could be
interpreted as "'schools don't make a difference,'" as educator Diane Ravitch
later wrote, certainly had dramatic policy implications, not least of all was
that "this finding raised serious doubts about the likely value of compensatory education for poor children, which was just beginning to burgeon in
response to the passage of federal aid to education only the year before."[65]
Small wonder, then, that the Johnson administration did everything it could
to ignore the report's publication, effectively shelving it so as to minimize
the political fallout.

At the same time that it did not do what it had been expected to do, the Coleman Report did represent a turning point in the intertwined histories of race and intelligence after *Brown*. The question of teacher expectation remained absent from the Coleman Report's analysis, although the report did rely on a "locus of control" framework. It had asked students whether they agreed or disagreed with statements like "Luck is more important than work" and "People like me don't have much of a chance." And it determined from students' responses that the minority child expressed the strongest "feelings of powerlessness" and imbibed an attitude that "outside forces over which he had no control would determine what kind of life he would lead." To substantiate this interpretation of its data, the report cited Lefcourt's research (which had, in turn, built on Rotter and Battle's study of schoolchildren) on a deficit in African American motivation as an explanation for poorer academic performance, making the observation that "for children from disadvantaged groups," a "sense of control of environment is most strongly related to achievement." It noted that for these children, "achievement or lack of achievement appears closely related to what they believe about their environment: whether they believe the environment will respond to reasonable efforts, or whether they believe it is instead merely random or immovable." The fact that the disadvantaged child routinely "experienced an unresponsive environment" meant, the Coleman Report added, that "the virtues of hard work, of diligent and extended effort toward achievement[,] appear to such a child unlikely to be rewarding" and that this in turn understandably led "to passivity" and "a general belief in luck, a belief that the world is hostile, and also a belief that nothing he could ever do would change things."[66] These remained key insights for James Coleman. He observed retrospectively in 1971, for instance, that a person's "orientation toward the environment" was a better gauge of that person's success in life than was his or her racial background; his report's data had demonstrated that the motivated African American teenager had on average far better verbal achievement test scores than did the unmotivated white teen.[67] By this circuitous path, it becomes evident how the Coleman Report's analysis (and its own adaptation of Lefcourt's version of Rotter's theory of locus of control) could have served to pave a way for Nixon and his administration to endorse the plasticity of intelligence while, simultaneously, taking that analysis to undercut rationales for the federal funding of compensatory programs to assist disadvantaged children.

A prime example of this double move appears in the way that Nixon's adviser on urban affairs, Daniel Patrick Moynihan, chose to read the Coleman

Report. Author of the contentious study *The Negro Family: The Case for National Action*, which had argued how the "dependence" of the African American family "on the mother's income undermines the position of the father and deprives the children of the kind of attention, particularly in school matters, which is now a standard feature of middle-class upbringing," Moynihan quickly became the Coleman Report's chief champion.[68] What Moynihan saw in the Coleman Report's data was a confirmation of his own prior suspicions about deficiencies in black families. In his view, the report underscored the significance of "the structural integrity of the home," and he saw in its data that "almost all groups show a tendency for achievement to decline in homes where the father is not present."[69]

Moynihan's reading of the Coleman Report was shared by the president. In his 1969 speeches, Nixon had expressly questioned the value of compensatory programming. By 1970, his staff was suggesting that school desegregation need not be pursued, either. In March of that year, John D. Ehrlichman, assistant to the president for domestic affairs, informally said to the press that he thought that "when a change in the racial makeup of the schools is undertaken for a purely social end, that's a misuse of the schools."[70] Later that same month, in a major address on school desegregation, Nixon spoke of the urgent "task of undoing the effects of racial isolation" and reaffirmed his "personal belief that the 1954 decision of the Supreme Court in *Brown v. Board of Education* was right." But in an apparent gesture to the Coleman Report (or at least to a Moynihan-sounding interpretation of it), Nixon went on to say, "The data tells us that in educational terms, the significant factor is not race but rather the educational environment in the home — and indeed, that the single most important educational factor in a school is the kind of home environment its pupils come from. As a general rule, children from families whose home environment encourages learning — whatever their race — are higher achievers; those from homes offering little encouragement are lower achievers."[71] In this way, Nixon made a progressive-sounding case about social environment to undercut a progressive argument for school desegregation — a move that was not so different from the one he had made the year before when he had spoken of the "learned helplessness" and the lack of "a sense of control" felt by disadvantaged children. It was, historians have observed, a "color-blind retreat" from school desegregation that sought to fragment support for school desegregation.[72] Indeed, this address has been cited as an excellent example of the "Southern Strategy" that brought Nixon to power in 1968 — a strategy that aimed to turn the American South from a

Democratic bastion into a Republican stronghold.[73] For when class status —
and not racial background — was announced as the key to a student's capacity to learn, why allocate enormous sums of federal funding to desegregate
schools? "There must be a better way to employ the material and political
resources of the federal government," Yale law professor Alexander Bickel —
an admirer of Nixon's position on education reform — pointedly observed on
the issue of school desegregation. "Massive school integration is not going to
be attained in this country very soon, in good part because no one is certain
it is worth the cost."[74] In a fascinating shell game, precisely the declared assertion of race's irrelevance in intellectual capacities created an opening for
racial prejudices to flourish once more.

Sociologist Gerald Grant has documented that Coleman never meant for
his report to fuel a backlash against either compensatory education for poor
children or efforts to further the integration of America's public schools. On
the contrary, Coleman and others instrumental in producing the *Equality of
Educational Opportunity* report emphasized in private interviews that they
had shared a fear that their report — in Grant's words —"would become a
weapon for racists" precisely because its findings suggested that "if Negro
and white achievement scores were different, but their schools weren't, didn't
that show that Negroes were intellectually inferior and that increased school
expenditures for them were a waste of money[?]"[75] Yet there were persistent
difficulties — both of data and of analysis, and the relationship between the
two — that Coleman could not resolve satisfactorily. In March 1970, Coleman
stated on the front page of the *New York Times* his own view that "school
integration is vital, not merely for some vague, generalized social purposes,
but because it is the most consistent mechanism for improving the quality
of education of disadvantaged children. If we abandon integration, we risk
creating all over the country the same kind of apartheid that has existed in
the South since Reconstruction."[76] But when he testified to Congress in April
1970 about the policy implications of his report, Coleman acknowledged that
his findings did support a conclusion that "the effects of changes in predominantly black schools, through compensatory programs . . . have been disap-
pointing. The addition of enormous resources (amounting in some cases to a
doubling of per pupil expenditures) in ghetto schools has not brought about
quality education or equality of opportunity."[77]

What had been a virtual consensus among leading educational psychologists that the achievement levels of disadvantaged children could be raised
through the combined efforts of early enrichment programs and school

desegregation turned out not to be enough to keep these agendas on course for more than a brief handful of years. After the publication of the Cole- man Report and its unanticipated conclusion that classroom quality and resources — and schooling more generally — mattered far less to children's academic achievements than did (depending on which findings were em- phasized) either their family background or their beliefs about their ability to master their own fates, prominent psychologists began to divide into op- posing camps on how best to interpret the report's findings. In the wake of the report, Harvard social psychologist Thomas Pettigrew argued that black children who possessed an internal sense of "'fate control'—indicated, for ex- ample, by disagreeing that 'Good luck is more important than hard work for success,'" did far better in school than black children who possessed an ex- ternal sense of fate control. However, Pettigrew too was wholly dismissive of "so-called 'compensatory education'" (as it would turn out, Rosenthal would be as well, albeit for different reasons). He noted that the "roughly billion- and-a-half dollars annually invested by the Federal Government" into these programs might be "politically expedient" because these efforts shifted the public's attention away from the "controversial and stoutly-resisted action" needed to force universal school desegregation. What funding for compen- satory programs, as opposed to a consistent commitment to desegregation, allowed was for the government simultaneously "to act and to avoid contro- versy."[78] Conversely, however, the Montessorian Joseph McVicker Hunt — long-standing advocate of early enrichment programs — observed in the same year, also with direct reference to the Coleman Report, that while it was "no wonder" that children "from the slums fail to persist in school- related tasks and tend to feel that circumstances control their fate," this proved precisely the point that desegregation without the supplement of compensatory programming would mean that "integrating the children of poverty in schools with children having had the advantages of middle-class background may compound the disadvantages of coming from poverty." In sharp contrast to Pettigrew, Hunt reiterated the vital role for "a large-scale effort to provide early education that would help to equalize the opportu- nities for children of the poor." The aim of compensatory projects, Hunt asserted, "is right, perfectly right."[79] The same evidence, in short, could be interpreted in completely different ways.

Finally, the Coleman Report could also be used to support an unabashed position that race was a biological category. By 1969 Berkeley educational psychologist Arthur R. Jensen, who would emerge as a dogged defender of a contrarian opinion that intelligence was *not* plastic and that African Americans were *never* likely to achieve educational equality, no matter the efforts made to "boost" their IQ, now declared that the Coleman Report provided excellent evidence for his own view that "environmental differences between groups is never sufficient in itself to infer a causal relationship to group differences in intelligence." For Jensen — here in stark opposition to Moynihan's take — the report demonstrated beyond a shadow of a doubt (at least for him) "that the factor of 'father absence' versus 'father presence' makes no independent contribution to variance in intelligence or scholastic achievement."[80] The Coleman Report served, however inadvertently, to fracture what had been a progressive psychological consensus, becoming a sort of looking glass into which social scientists came to see their own perspectives on the educability of the disadvantaged child reflected back at them.

THE COLEMAN REPORT'S unexpected findings were being debated in a political landscape decidedly different from the one in which the report had been first proposed in 1964. Compensatory education programs to improve the intelligence of disadvantaged students were — after previously appearing to show documentable gains — suddenly failing to yield encouraging evidence. An evaluation of Harlem's Higher Horizons found the project had "virtually no measurable effect in terms of educational achievement," and this led in 1966 to a decision by the New York City school system to shut it down.[81] In 1968, Edmund Gordon and Adelaide Jablonsky, writing in the *Journal of Negro Education*, reported that achievement gains among Head Start children turned out to be "equivocal" and that compensatory education efforts funded by the Elementary and Secondary Education Act had "not yet resulted in a major change in the schools' success patterns with children from disadvantaged backgrounds."[82] In 1969, a study conducted by the Westinghouse Learning Corporation and Ohio University concluded "that Head Start as it is presently constituted has not provided widespread significant cognitive and affective gains which are supported, reinforced, or maintained in conventional education programs in the primary grades."[83] It had been to this still-unpublished study that President Nixon was alluding in his speech on April 9, 1969.

Few could have anticipated that the Coleman Report might do damage to the legacy of the *Brown* decision and the cause of school desegregation.[84] Yet a Harlem schoolteacher writing in late 1967, just as Nixon was preparing to assume the presidency, predicted in *Integrated Education* — a journal whose advisory board included a host of well-known educational psychologists, sociologists, and educators — that, "as a whole, the Coleman Report, despite its incidental usefulness in documenting certain aspects of student and teacher attitudes and in compiling data and correlations on student achievement tests, will prove a disservice to those interested in both quality and integrated education." The teacher, Deborah W. Meier, elaborated eloquently, "It will do so because it not only asks the wrong questions but asks them in the wrong way, and even more important, because it avoids examining and exploring certain facts about our school system and its social setting, facts which are essential to an understanding of what is going on." Meier took special exception to the report's view that there existed a "relationship between a student's achievement in school and his feeling that he had 'control' over his own future." She continued, "Learning occurs best when students, parents, and I might add teachers, have not merely a 'sense of' control but some real control over their destiny. The 'feeling of' control, however, can best be fostered when young men and women can reasonably look forward to decent jobs, good housing, and a world of peace and mutual respect. In the short run, within the school, it can only be sustained by increasing the control which students, teachers, and community have over all areas of their joint educational experience."[85] Meier — later to become the celebrated author of such texts as *The Power of Their Ideas* and *In Schools We Trust* — was one of a very few who saw that the terms of debate within which the Coleman Report was framed had been fatally flawed from the start.

How *Pygmalion* Fails

The shift in political climate was manifest. Initially Rosenthal and Jacobson's book won plaudits for its argument about teacher expectancy effects. Its defense of the malleability of children's intelligence had already earned *Pygmalion in the Classroom* recognition on the *New York Times'* front page nearly a year before the book's publication.[86] Rosenthal and Jacobson's research was widely interpreted as an attack on the ways in which the "ability grouping" of students served to reinforce inequities in education. It fueled efforts to reduce the tracking that was becoming an increasingly standard practice in both elementary and secondary schools — as well as a practice that saw African American students "more likely to be in the lowest track and less likely to be in the highest," as the Coleman Report had concluded.[87]

The decision in the *Hobson v. Hansen* federal court case (filed by a civil rights activist against the District of Columbia's board of education) centrally cited Rosenthal and Jacobson's research when the court ruled in 1967 to eliminate "ability grouping" (due to racial bias).[88] Rosenthal and Jacobson's research got tangled up as well in the quite heated debate over community control of school boards — an issue that resulted in the shutdown of New York's public schools in 1968.[89] And although Rosenthal and Jacobson had relied on IQ testing to demonstrate the effectiveness of their approach, their work was additionally used to critique an overreliance of school systems on intelligence testing as a means of measuring scholastic ability; citing *Pygmalion in the Classroom*, the board of education in Los Angeles moved to ban IQ testing of all its elementary school children.[90]

Yet the book also contributed directly to opposition to federally funded compensatory education programs. Strikingly, Rosenthal and Jacobson — whether in sincere belief or in strategic sync with the trends of the time, or a combination of both — sided fully with the view that compensatory education represented a waste of time and money. In *Pygmalion in the Classroom*, they made clear their broadly dismissive attitude toward "expensive, special programs" designed "to overcome learning handicaps" of the "disadvantaged child" by "means of acting on the child — remedial, reading, counseling, and guidance, cultural experiences, parental involvement, and health and welfare services." For Rosenthal and Jacobson, it was a serious limitation of federal educational programs that they were basically "constructed to emphasize deficiencies within the child and the home, and they are all compensatory approaches." All "too rarely"— and here they restated their core position — did these programs entertain the notion "that teacher attitudes and behavior might be contributing factors to pupil failure."[91] Taking their case to the public not only in their book but also in the pages of *Scientific American*, Rosenthal and Jacobson elaborated still further their opposition to federally funded early education programs, arguing that a teacher's "tone of voice, facial expression, touch and posture may be the means by which — probably quite unwittingly — she communicates her expectations to the pupils." Here too they explicitly endorsed the view that remedial programs were turning out to be ineffective. "For almost three years, the nation's schools have had access to substantial Federal funds under the Elementary and Secondary Education Act, which President Johnson signed in April 1965," they wrote. "The premise seems to be that the deficiencies are all in the child and in the environment from which he comes." Rosenthal and Jacobson sharply contrasted their own psychological research with these federal efforts: "In our experiment nothing was done directly for the child. There was no crash program to improve his reading ability, no extra time for tutoring, no programs of trips to museums and art galleries. The only people affected directly were the teachers; the effect on the children was indirect." And yet the gains in IQ as a result of ESEA-funded "remedial instruction, cultural enrichment and the like," they declared, turned out to be "far smaller than the gains made by the children in our experimental group." Rosenthal and Jacobson asked, "Have the schools failed the children by anticipating their poor performance and thus in effect teaching them to fail?" Or putting their case more strongly still, "Are the massive public programs of educational assistance to such children reinforcing the assumption that they are likely to fail?"[92] Here then was another (self-understood as liberal) argument that, at one and the same time, denigrated federal efforts geared to intervening in the lives of disadvantaged schoolchildren *and* neglected to call for the full racial integration of American schools. Its perspective was humanistic. But it was oddly remiss in not acknowledging the possible role that larger structural problems might play in the achievement potential of disadvantaged youth. Apparently, the way to help children most effectively was to focus on one classroom teacher at a time. An enormous expenditure of federal funds was never going to make as much of a difference.

However, the decision by Rosenthal and Jacobson to rely so narrowly on cognitive testing as the sine qua non measure of a student's success quickly got turned against them. Critics questioned whether there really was any such thing as a Pygmalion effect in the American classroom. Several educational psychologists went on the offensive against *Pygmalion in the Classroom*. Robert L. Thorndike angrily noted how the analysis of Rosenthal and Jacobson was "so defective technically that one can only regret that it ever got beyond the eyes of the original investigators!"[93] Lee J. Cronbach concluded that "in my view, *Pygmalion* merits no consideration as research," adding as well that "the advertised gains" of the children in the book were quite likely "an artifact of crude experimental design and improper statistical analysis."[94] Nor did it help *Pygmalion*'s case when the teacher-focused Banneker Project in St. Louis came reluctantly to acknowledge that the academic achievements of its African American students turned out to be no different from those of students not part of the project.[95] And neither was its cause aided when the *Journal of Educational Psychology* announced that its several attempts to replicate Rosenthal and Jacobson's experiment had resulted only in failures.[96]

More devastatingly still, Berkeley psychologist Jensen quickly seized on the many emerging criticisms of *Pygmalion*'s methodology to make a case that Rosenthal and Jacobson's findings actually reinforced a view that genetics determined a child's intelligence. While a decision to use IQ as their key metric was standard in that era, as historian of education Barbara Beatty has observed (although without any reference to *Pygmalion*), it came with its own set of attendant risks: "Proving that preschool education could raise IQ was part of the long quest to refute genetic and racial arguments about intelligence, but using IQ scores as a measure of success also provided fodder for critics like Jensen when gains faded."[97] *Pygmalion*'s reliance on IQ was no exception; it left Rosenthal and Jacobson wide open to Jensen's attack. Jensen ridiculed the notion that "a 'stimulating' environment" alone could ever reverse the principal role played by "hereditary" and "genetic" factors in the formation of a child's intelligence. As he wrote, "The belief in the almost infinite plasticity of intellect, the ostrichlike denial of biological factors in individual differences, and the slighting of the role of genetics in the study of intelligence can only hinder investigation and understanding of the conditions, processes, and limits through which the social environment influences human behavior."[98] Specifically with respect to *Pygmalion*, Jensen scathingly stated, "Because of the questionable statistical significance of the results of this study, there may actually be no phenomenon that needs to be explained."[99] The very cacophony of the discussions of possible factors that might explain the academic difficulties of "the disadvantaged child" had the unforeseen outcome of consolidating a conviction that this child's levels of achievement and motivation could not be helped.

Conclusion

In 1964, Martin Seligman was a new graduate student in behavioral psychology at the University of Pennsylvania when he first witnessed a series of conditioned response experiments conducted with dogs. The classic Pavlovian procedure was to place a dog in one half of a "shuttle box," a rectangular cage with two equal compartments divided by a barrier. The dog then received an unpleasant stimulus (typically an electric shock) that prompted the animal to jump across the barrier to the other half of the apparatus (where there was no shock). The dog in this way learned to escape pain. Psychologists called this behavioral model "avoidance learning," and the shuttle box had been successfully used for some decades to demonstrate this experimental paradigm in animals. But the avoidance learning experiment at Penn was intended as a new variation on this old theme.

The experiment Seligman witnessed had a dog strapped into a harness while researchers applied electric shocks to the dog's back paws. It then involved allowing a day and a night to elapse before experimenters placed the dog in the shuttle box. After being placed in the shuttle box, the dog heard a distinctive tone several seconds before it subsequently received an electric shock. The hypothesis was that the dog could learn to respond to the tone, grasping that it represented an early warning signal that pain was imminent; the dog would then jump across the barrier at the sound of the tone, having learned to avoid the shock entirely. But in the experiment the dog did not learn to avoid the pain; it did not jump away from the electric shocks. A dog initially placed in circumstances where it had to endure an unavoidable painful electric shock later became passive and seemingly unable to escape the shock even when it had both the warning cue and the physical opportunity to do so.[100] The experiment was classed a failure.

Yet Seligman, along with his fellow graduate students Steven Maier and J. Bruce Overmier, recognized the prospect of another behavioral paradigm altogether. Seligman and Maier recalled many years later, "We thought that a profound failure to escape was *the* phenomenon and we began to try to understand it."[101] And so they conducted a series of animal experiments that took this puzzle of the seemingly fatalistic dog and arrived at their own insight: that animals first conditioned to accept an uncontrollable shock would later become unable to avoid the shocks. Due to the initial inescapable nature of their circumstances, dogs learned that resistance to the painful shocks was futile. They howled when hit by the electric shocks, but they still made no attempt to escape the shocks. Repeated trials only made matters worse; the dogs did not learn to escape. Instead they grew increasingly unreceptive when hit by the electric shocks. Efforts were made to motivate the increasingly passive dogs to rescue themselves from traumatic electric shocks. The barrier between the two sides of the box was removed; meat was put on the nonelectric side of the box; the dogs were called to cross over to the non-electrified side of the box. "Nothing worked," Seligman wrote in 1969. "As a last resort, we pulled them back and forth across the box on leashes, forcibly demonstrating to them that movement in a certain direction ended shocks.

That did the trick, but only after much dragging."[102] The helplessness of these traumatized dogs, having been learned, turned out to be intensely difficult (though not impossible) to unlearn. It was as if the dogs had simply given up.

Seligman, Maier, and Overmier proposed that the dogs had come to suffer from a condition they named "learned helplessness," defined as "the failure to escape shock by uncontrollable aversive events." "Learned helplessness might well result from receiving aversive stimuli in a situation in which all instrumental responses or attempts to respond occur in the presence of the aversive stimuli and are of no avail in eliminating or reducing the severity of the trauma," concluded Overmier and Seligman in 1967.[103] In 1968, additionally, Seligman and Maier, together with psychologist James H. Geer, went on to link their findings to Lefcourt's work on locus of control, noting that "the perception of degree of control over the events in one's life seems to be an important determinant of the behavior of human beings."[104] (Seligman would subsequently acknowledge his debt to the precursor research of Julian Rotter.)[105]

Still, and despite Nixon's reference in 1969 to the chronic damages done by learned helplessness in impoverished children, no actual research had been conducted on impoverished children and learned helplessness. There were general claims that learned helplessness *might* likely be applicable to people, but no experiments had been conducted to test such claims. In 1968, for instance, Seligman, Maier, and Geer proposed that their research on "the maladaptive failure of dogs to escape shock" bore a strong resemblance to "some human behavior disorders in which individuals passively accept aversive events without attempting to resist or escape." Here they cited as proof of learned helplessness in humans the reflections of psychologist Bruno Bettelheim that Nazi concentration camps at Dachau and Buchenwald (where Bettelheim had been an inmate) had transformed some of their prisoners into "walking corpses" who had opted passively not to struggle to remain alive.[106] But as late as 1972 Seligman still felt obligated to acknowledge, "Since most of the investigations of uncontrollable trauma have used infra-humans, we can only speculate on the relationship of learned helplessness in animals to maladaptive behaviors in man."[107] To put it another way: Nixon's claim that learned helplessness theory could be applied to human beings preceded by several years the empirical evidence that this was so. (It would not be until 1973 that developmental psychologist Carol S. Dweck — although with no attention to race or class — published the first results of a series of clinical experiments that sought to explore the susceptibility of schoolchildren to feelings of learned helplessness and the potential consequences of these feelings for their academic performance.)[108]

It was Seligman himself who, albeit with a remarkable degree of awkwardness, attempted to bring plasticity of intelligence, education reform, and the dynamics of race and class directly into discussions surrounding learned helplessness. In *Helplessness*, his book-length study of 1975, he lingered at some length on "methods of curing helplessness and of preventing" the condition, as well as on the link between learned helplessness and "the cognitive development of a child." Here Seligman briefly invoked a *Pygmalion*-sounding analysis when he noted how "with a certain teacher or with a certain subject, the child may feel helpless."[109] But for the most part, it was as if Seligman's views on the disadvantaged child were meant to pay homage at long last to Richard Nixon for having brought his psychological term to national prominence.

In the book, Seligman addressed how so "many aspects of poverty converge in their effects by producing helplessness." In speculating, for example, how poverty might lead to learned helplessness in the young, he imagined this nightmarish scenario: "Extreme, grinding poverty does produce helplessness, and it is a rare individual who can maintain a sense of mastery in the face of it. A child reared in such poverty will be exposed to a vast amount of uncontrollability. When he cries to have his diaper changed, his mother may not be there — or if there, too exhausted or harried to react. When he is hungry and asks for food, he may be ignored or even struck. In school, he often will find himself far behind, bewildered, and even abused." Seligman further asserted that a theory of helplessness had significant implications for education reform, and he announced that learned helplessness might indeed explain many cognitive difficulties in schoolchildren. "What is often passed off as retardation or an IQ deficit may be the result of learned helplessness," Seligman contended. "Intelligence, no matter how high, cannot manifest itself if the child believes that his own actions will have no effect." Yet at the same time, Seligman made a variety of suggestions to the effect that social policies designed to assist the poor were counterproductive. "The welfare system, however well intentioned, adds to the uncontrollability engendered by poverty," he offered (albeit with no supporting evidence). "It is an institution that undermines the dignity of its recipients because their actions do not produce their source of livelihood."[110]

Seligman did push back against an argument that the IQs of black American children were lower than the IQs of white children due to genetic differences between the races, and here too Seligman used helplessness theory to advance an alternative explanation. Since feelings of helplessness caused depression and depression diminished cognitive capacity, Seligman wrote, there existed a strong possibility that the "motivational impairment" produced by depression, "rather than 'intellectual' inferiority," could well be a decisive factor in the differential IQ scores between white and black children. "No study that I know of has ruled out such a belief in helplessness as a cause of the lower IQ scores and scholastic performance of poor black American children," he concluded. Yet even as he distanced himself from a Jensen-style argument that IQ was principally biological, Seligman nonetheless concluded frankly that his own research represented "no exception to the trend away from plasticity." Seligman announced regrets that he now accepted that "the cognitive development of a child is not nearly as plastic as I had hoped," adding that he was "no longer convinced that special, intensive training will raise a child's IQ by twenty points, or allow him to talk three months early, or induce him to write piano sonatas at age five, as Mozart did."[111]

What had begun approximately a decade before as a dedicated movement radically to reform the American school system with the extraordinary assistance provided by the Elementary and Secondary Education Act of 1965 — a movement undergirded by the passionate arguments of educational and developmental psychologists committed to a belief in the plasticity of intelligence — had been pushed back by the time of Seligman's book. The ensuing discussion was ideologically muddled. Through the 1970s, learned helplessness became a new reference point in analyses of race, class, and education in America, blurred together with (or used instead of) a concept of locus of control.[112] At the same time, the idea of plasticity of intelligence receded; a child's failure or success came far more centrally to be understood as related to self-concept — most fully dependent on what the child *felt* or *imagined* his or her own chances at academic achievement to be.

TWO

Minimal Brain Dysfunction, Ritalin, and Racial Politics

O N JUNE 29, 1970, a front-page article in the *Washington Post* offered the shocking statistic that between 5 and 10 percent of all schoolchildren in Omaha, Nebraska, were receiving "'behavior modification' drugs prescribed by local doctors to improve classroom deportment and increase learning potential." The article went on state that stimulant drugs like methylphenidate (better known as Ritalin) were being given to "'hyper-active students'" as part of a citywide program known as STAAR (an acronym for Skills, Technique, Academic Accomplishment, and Remediation) and that this new program had received the blessing — if not the outright endorsement — of the Omaha public school system. The idea behind STAAR, as pediatrician Byron B. Oberst — who had first proposed the plan to Omaha school administrators — informed the *Washington Post*, was that "we know these children become more successful" when they are taking stimulant drugs. "They become more self-confident," Oberst added. As for specific evidence that a program that provided Ritalin to large numbers of Omaha schoolchildren had been proven effective, Oberst referred to his own young patients who were demonstrating "marked improvement in handwriting and fine motor co-ordination problems."[1] However, the real drama in Omaha was not whether stimulant drugs turned fidgety, underperforming students into well-behaved, competent pupils, the *Post* article continued, but whether the schoolchildren singled out by teachers and school officials to receive stimulant medications were disproportionately poor and nonwhite.

Anecdotal evidence strongly suggested that they were. Maryl J. Harris, a black militant spokesperson, declared as much at a local school board meeting two weeks before the *Post* article appeared. She claimed that teachers had been dispensing Ritalin to children at the predominantly African American elementary school in Omaha (where Harris taught a black studies course) only in order "to slow kids down so they can control them instead of teach them."[2] Ernest W. Chambers, an African American candidate for the Nebraska state legislature, made the Omaha program of prescribing stimulant drugs to students a signature campaign issue (see figure 3). Chambers asserted provocatively in a position paper, "'Speed' for Slow Learners — Drug Tests on Omaha School Children," that "ghetto children are placed on drugs because their 'behavior' happens not to please some teachers or principal," adding that the real intent of "the Omaha experiment" (as it came to be known) was to "dope" the city's black youth "into submissive zombies."[3] (Chambers went on to win the general election.) Others in the African American community made similar statements to the press; a black mother in Omaha told *Newsweek*, for instance, that her son's teacher had "badgered us for a month and a half" until the teacher had coerced her to sign a consent form that allowed the school to prescribe the boy "pep pills."[4] The pediatrician who had introduced the idea, Byron Oberst, did little to quell the growing controversy. On the contrary, when Oberst spoke to the *Washington Post*, he may have only inflamed public opinion further. "I would certainly hope," Oberst said of Omaha's stimulant drug program, "that this would be beneficial to the low income family." To which the *Post* noted with a hint of the Orwellian, "No limit seems to have been reached yet in the application of drugs to the problems of behavior."[5]

Did the Omaha experiment have a disturbing racial or class agenda? African American spokespersons and their allies alleged that stimulant drugs were being pressed upon far too many black and poor schoolchildren with far too little oversight. Activists observed that black children were getting Ritalin prescriptions for a behavioral disorder that had no name, no reliable diagnosis, and no clear set of symptoms. Was it minimal brain dysfunction (MBD), hyperkinesis, hyperactivity, or cerebral dysfunction? Why was it that negative electroencephalogram results "did not rule out minimal brain dysfunction"?[6] How was it "that even when brain damage could not be demonstrated it could be presumed to be present"?[7] Why would a black and poor child who was "continually in motion, cannot concentrate for more than a moment, acts and speaks on impulse, [and] is impatient and easily upset" be determined to have MBD — and be a prime candidate for a Ritalin prescription — since so many children of all colors and classes exhibited

FIGURE 3. Ernest Chambers addresses the congregation in Omaha's Calvin Memorial Presbyterian Church in 1968. Four years earlier, Chambers had met Malcolm X during a visit to Omaha. Two years later, Chambers would be elected to the Nebraska state legislature to represent North Omaha's Eleventh District. As state representative, Chambers successfully campaigned for a bill that enabled African Americans to be elected for the first time to the Omaha city council and school board. Chambers later won support for a resolution that called on Nebraska to divest from South African apartheid—making it the first state to do so. Photo by Bettman via Getty Images.

some of these behaviors at some point?[8] Many black parents expressed anger and distrust at such professional determinations, and their protests against the Omaha plan (and soon against stimulant drug plans in other school systems around the country) set in motion a fierce national debate about potential abuses surrounding the prescription of stimulant drugs to children.

Yet, and despite the seeming centrality of race to the intertwined histories of hyperactivity and stimulant drug use with schoolchildren, the politics of race has not been given its due in accounts of these subjects. The protests in Omaha and elsewhere receive routine mention, but there remains no genuine analysis of the roles played by racial politics in the ongoing debates around stimulant drugs and hyperactive children.[9] We are told that the allegations against the Omaha school systems were largely false and exaggerated. It was not true, for instance, that the numbers of minority children who were receiving stimulant drugs were as high as initially stated; the percentage of children prescribed Ritalin in Omaha was far lower than the 5 to 10 percent cited in the *Post* article.[10] It was also not true that the parents of children selected for Ritalin prescriptions were generally coerced to take stimulant drugs. Rather, the historical record indicates that parents were usually the ones who initiated a request for a medical referral for their children.[11] Recent studies have additionally noted "mounting evidence that children from lower socioeconomic statuses, as well as minorities such as blacks and Hispanics, are 'less likely to have the diagnosis even after controlling for other characteristics.'"[12] Therefore, and fascinatingly, when the topic of race has made an appearance, however cursorily, in histories of hyperactivity and stimulant drug use with children, the conclusion has been that children of color were *not* preselected for stimulant drug programs in the course of the 1960s and 1970s — or in the several decades since — but rather that *white* children were widely *overrepresented* when it came to these programs. Giving methylphenidate to schoolchildren quite quickly became a white and middle-class thing — another detail of the story that has been gestured to in passing but not addressed or analyzed.[13]

Important historical questions, then, remain unasked and unanswered. How did commentators and activists at the time manage to get their stories about the racialized use of stimulant drugs in children so wrong? How did a politics of race get enmeshed with the histories of stimulant drugs and hyperactivity in the first place — and when and how did misinterpretations of the interrelationships between stimulant drugs, hyperactive children, and racial politics emerge? Why did this misinterpretation persist for so long despite empirical evidence that challenged it? And perhaps most important, why did white children become more likely to receive diagnoses for hyperactivity — and stimulant drug treatment — than children of color? As this chapter aims to demonstrate, answers to these questions have much to do with the history of the field of psychology, for it was psychologists who played a central — if often ideologically ambiguous — role in the course of the 1960s and 1970s both in the initial identifying and in the subsequent diagnosing of the disorder in children that came to be known as hyperkinesis or minimal brain dysfunction. The collaboration of psychologists with child psychiatrists, pediatricians, neurologists, and other physicians became essential to clinical assessments of the effectiveness of potential drug treatments for this — asserted to be — widespread behavioral condition in children.

The Vanishing Point of Race

Before it became a major focus for controversy in the early 1970s, the racial background of the children given stimulant drugs was seldom discussed. Yet it was black youth on whom stimulant drugs were initially tested. The pioneering psychopharmacologic research of Leon Eisenberg represented a prime example. In the early 1960s Eisenberg was a child psychiatrist at Johns Hopkins University School of Medicine, with a long history of caring for troubled minority youth. At that time, Eisenberg served as a psychiatric consultant for emotionally disturbed foster children referred to him by the urban welfare department at Baltimore City Hospital. The children were disproportionately African American (over 60 percent) in a city that was approximately one-third black; most were boys. In 1961 Eisenberg had delivered a passionate address to the annual meeting of the American Orthopsychiatric Association that lamented how these "children of the lower depths, children mistreated before they came into care and treated not too well afterward," were caught up in the bureaucracy of "a public agency, which in the United States means it is inadequately funded, constantly beleaguered by ignorant critics, fighting to maintain itself marginally." Eisenberg called on fellow "members of the orthopsychiatric professions" to acknowledge what he saw as their "heavy moral responsibility" to work to "provide leadership for community action to correct social and institutional hazards that impair the physical and mental health of these children."[14] It was not long afterward that Eisenberg began to publish the results of his experiments treating this same cohort of children with stimulant medications.

In 1962 at the annual meeting of the American Orthopsychiatric Association, Eisenberg and pediatric psychologist C. Keith Conners (along with several associates) presented the initial results of a stimulant drug trial conducted with minority children. Their subjects were all African American boys (ages seven to fourteen) institutionalized at a Maryland training school for delinquent youth (formerly known as the House of Reformation and Instruction for Colored Children). The boys were given the stimulant medication dextroamphetamine and then observed for alterations in their behavior.

However, the results were inconclusive; the experimental design had been flawed. That the children lived together in cottages, that they knew they were receiving medication, and that the experiment had been disrupted by a brawl between staff and subjects were all considered to have thrown off results. Eisenberg concluded with a socioeconomic — not a pharmacologic — observation, noting that "drugs will not in any way diminish the necessity for more and better trained personnel and well-conceived programs of rehabilitation if any advantage is to be taken of the amelioration of behavior produced by medication."[15] He persisted in his conviction that *social* factors should remain central to any discussion of stimulant drug treatments with troubled youth.

By 1963, Conners became the lead author with Eisenberg on a (now canonical) study that appeared in the *American Journal of Psychiatry*. This study also used emotionally disturbed children in foster care as its subjects for stimulant drug treatment. But these subjects included both boys and girls (though twice as many boys as girls) in the age range from seven to fifteen. And these subjects were mainly white children (more than 80 percent were white, while the remainder was African American). Conners and Eisenberg reported these variables but did not discuss them. As for the trials, these were now double-blind. Some children unknowingly got a placebo; others unwittingly received methylphenidate. Neither Conners and Eisenberg nor the children's caretakers knew which child was receiving medication (although Conners and Eisenberg suspected that the caretakers may have guessed due to the marked side-effect of loss of appetite noted among those receiving methylphenidate). When Conners and Eisenberg asked the children to describe their moods, the ones subsequently identified as having received medication described "'feeling happier.'" And when Conners and Eisenberg conducted several intelligence tests with the children, they found that those treated with drugs demonstrated "significantly better performance," an indication that Ritalin had "an interaction with I.Q."[16] Historians of medicine observe that this article by Conners and Eisenberg became "the first article in a medical journal explicitly advocating the use of Ritalin" for treatment with children.[17] However, it has not been noted how race went in and out of focus in this article — at once mentioned but never discussed. And when Conners and Eisenberg published a follow-up double-blind study in 1964 on the effects of methylphenidate on a comparable cohort of children ("wards of the city who are awaiting their first foster home or who have failed to adjust in previous foster homes"), here the race of the subjects dropped away entirely — even as these trials determined methylphenidate to have only an *insignificant* effect on these children's ability to learn.[18]

It was not that Conners and Eisenberg ceased to concern themselves with the racial background or class status of their drug trial subjects. On the contrary, the evidence suggests that they continued to care deeply about the variables of race and class. It was only that they considered these variables to be essentially irrelevant in terms of how a particular child might respond to a particular stimulant drug treatment. Any difference in the way a child responded to Ritalin was going to be an individual difference, especially given that (as they wrote in 1964) "the mechanism of action of methylphenidate is still not known." As they additionally stated, it was entirely possible that gains witnessed in drug-test subjects were the results of a "strong desire" on the part of subjects "to reward the experimenters with a good performance" and therefore had next to nothing to do with the effects of "pharmacology."[19] By 1967, Conners and Eisenberg furthered this analysis in a stimulant drug treatment study conducted with African American children with behavioral disabilities, concluding, "Despite the finding of rated improvement in three areas of classroom behavior, attitude toward authority, and group participation, it seems doubtful that the medication specifically affects these areas."[20] Here they again expressed their hesitation as to whether stimulant drug treatments could be used as a means to improve a student's performance.

There is no evidence that Conners and Eisenberg came in the course of the 1960s to consider the minority child a more likely candidate for stimulant drug therapy. Eisenberg and Conners may have conducted their first stimulant drug trials with children of color, but when asked in 1966 to evaluate a six-week Head Start program's impact on the cognitive development of several hundred "severely disadvantaged" kindergarten children, they said nothing at all about stimulant drugs. They focused on the "remarkable" gains made by the Head Start children; they worked hard to applaud Head Start for what it stood to achieve and emphasized only the social dimensions of the problems confronting these children. But they also underscored that a short-term fix was unlikely to repair a long-term and systemic problem. As Eisenberg and Conners wrote, they were "far from convinced that these [children's] gains will endure, given the over-crowding, educational impoverishment, and generally negative attitudes toward the poor that characterized inner-city elementary schools."[21] In this way, they consistently indicated that the hurdles faced by poor and minority children were principally social and economic — and that any lasting solution to these hurdles should involve social and economic change.[22]

In late 1970 Conners published the results of a study he had likely completed before the Omaha experiment became national news. Here he reported on the "symptom patterns" of more than 350 "hyperkinetic" children at four public schools — one white and middle-class, a second white and lower-class, a third African American and middle-class, and a fourth African American and lower-class. Lower-class children demonstrated marginally higher rates of antisocial behavior than did middle-class students; black students scored higher in the category of "problems in school" than did white students (who scored higher in the category of "overasserts self"). Conners wryly observed, "It would be hazardous to infer that the actual symptom *rates* among lower-class Negro children are different. At most, one can only conclude that lower-class Negro mothers respond differently than their middle-class counterparts to the questionnaire." Conners's results revealed little of statistical significance; the effects of race and class were not measurably evident when it came to the symptoms of hyperkinesis. "It is perhaps well to remember the dictum that the first thing most questionnaires measure is social class," he noted.[23] Nonetheless and despite that assertion, Conners found no meaningful differences in the symptomology of children with hyperkinesis based on either racial background or socioeconomic status. Conners emphasized in 1972, "All children with this diagnosis [of minimal brain dysfunction] do not respond in the same way to drug therapy."[24] The success or failure of a particular drug treatment depended on the specific physiological and psychological challenges confronted by an individual child — not on the race or class background of that child.

In the early 1970s, Eisenberg more explicitly expressed frustration at how debates over the use — and abuse — of stimulant drugs to treat poor and minority children diagnosed with hyperactivity had evolved. He voiced his impatience at those who characterized treatment with stimulant medication as "a nationwide conspiracy to drug children into insensibility," observing that the idea "that stimulants make hyperkinetic children into conforming robots is arrant nonsense."[25] Yet Eisenberg fully acknowledged the potential for gross abuses, especially when it came to poor and minority schoolchildren, noting not only that "restlessness in an inner city classroom can be diminished by feeding the children since many come to school without breakfast and exhibit restlessness resulting from hypoglycemia" but also that "it would be a criminal abdication of medical responsibility to treat such children with stimulant drugs."[26]

Throughout the 1960s and into the 1970s, then, Conners and Eisenberg accepted as an article of faith that the question of whether a stimulant medication like methylphenidate would likely assist an individual child with his or her behavioral problems could not be correlated with that child's racial or socioeconomic background. They did not start from a premise that poor and minority children were more likely to benefit from methylphenidate than were children who were white or middle-class — or both. Despite their advocacy of stimulant drug treatment under some circumstances for some children, Conners and Eisenberg maintained a healthy skepticism concerning the prospect that methylphenidate possessed more universal curative properties.

White Children and Minimal Brain Dysfunction

It had been the Rhode Island pediatrician Charles Bradley who first reported in 1937 that amphetamines were helpful "as a therapeutic agent for maladjusted children." After administering the stimulant drugs Benzedrine and Dexedrine, Bradley observed that some (though not all) of the children began "to walk and move quietly in contrast to previous noisy running and rushing about. A number spoke in normal or lowered tones of voice instead of shouting raucously."[27] He also wrote that "it appears paradoxical that a drug known to be a stimulant should produce subdued behavior in half of the children," while at the same time "possibly the most spectacular change in behavior brought about by the use of benzedrine was the remarkably improved school performance of approximately half the children" treated with stimulants.[28] In 1957 psychiatrists Maurice Laufer and Eric Denhoff exponentially expanded the category of children who were likely to be afflicted with hyperactivity — and therefore might be assisted by the administering of amphetamines. Their work emphasized that hyperactivity represented a "common behavior disorder in children"— including among children of "normal intelligence" when their "involuntary and constant overactivity" greatly "surpasses the normal."[29] That no measurable brain damage could be found in these children did not mean that they did not suffer from a brain dysfunction. Rather, observation of their behavior could be used to infer that they *did* have a central nervous disorder. That a child was determined to be hyperkinetic became a key indicator of that child's disorder — in the absence of conclusive organic evidence.

Laufer and Denhoff introduced a new diagnostic category: hyperkinetic impulse disorder. And this new diagnostic classification, as historian of medicine Matthew Smith has written, became "a point of departure for modern conceptions of hyperactivity by depicting the disorder as one that could be applied to millions of children."[30] In the course of the 1960s, diagnoses of hyperactivity and hyperkinesis in school-age children were seen to grow to near-epidemic proportions. Already in 1962, or a mere five years after Laufer and Denhoff published their results, hyperactivity had already become regarded as "one of the most common behavior disturbances observed in children."[31] By the mid-1960s, a concept of "the hyperactive child" had become well established in the pediatric literature, even as the behavior of this child remained terribly broad; "can't sit still," "talks too much," "gets into things," "doesn't stay with games," "leaves class without permission," and "unpopular with peers" were some of the more prominent symptoms. Strikingly, pediatricians began additionally to note that "several of the mothers of our patients have described their children as having 'Jekyll and Hyde' personalities," a characterization that would recur in the literature on hyperactivity.[32] Children most at risk of hyperactivity were invariably said to be "of normal intelligence."[33]

The decade of the 1960s witnessed the rapidly accelerated promotion of another deliberately vague diagnostic term: minimal brain dysfunction. Sam D. Clements, a child psychologist based at the University of Arkansas's medical center, played a central role in this regard. Clements was appointed in 1963 as the project director of a special task force established by the U.S. Department of Health, Education, and Welfare to investigate the educational consequences of "deviations in nervous system function" on untold numbers of American schoolchildren. In 1966, Clements published the first report from the special task force; in it he argued that there should be no requirement that children with learning disabilities demonstrate "physiologic, biochemical, or structural alterations of the brain." Clements went on: "We cannot afford the luxury of waiting until causes can be unquestionably established by techniques yet to be developed." And in this context Clements proposed that "minimal brain dysfunction" become the preferred term for this new nonspecific condition — precisely because the adjective "minimal" allowed "milder, borderline, or subclinical abnormal manifestations of motor, sensory, or intellectual function" to be included in diagnoses along with perceived deficiencies in "learning, thinking, and behavioral sequelae." The classification of "minimal brain dysfunction" meant, Clements wrote, that children whose "symptomatology" included "near average, average, or above average general intelligence with certain learning or behavioral disabilities ranging from mild to severe, which are associated with deviations of function of the central nervous system," could be differentiated from children with more severe disabilities like cerebral palsy and epilepsy.[34] It also meant that children diagnosed with MBD were far more likely to be white.

Clements did not state this final point quite so bluntly. Rather, he wrote that MBD should *not* be read as an attempt to seek federal funds to assist in serving the educational needs of minority or underprivileged children. The introduction of MBD into policy discussions, Clements said, was not part of an effort to generate compensatory programming for children whose environments were said to impede their ability to learn. In 1966 Clements wrote that "the evaluation of the intellectual functioning of the 'culturally disadvantaged' child, though perhaps related, represents an equally complex, but different problem" from that of the child with minimal brain dysfunction.[35] In short, MBD was not intended — or invented — as a diagnosis to be principally applied to black or poor children. Minimal brain dysfunction was to be a condition found in the children of the middle classes (often simply presumed by various commentators in discussions of MBD to be white).

In this way, MBD stood in contrast to yet one further diagnostic classification advanced at this same moment: mild mental retardation. Historian Mical Raz has documented that a diagnosis of mild mental retardation became quickly identified in the 1960s with low-income and minority children. The liberal rationale behind this diagnosis, Raz writes, was that minority and poor children in inner-city communities were being multiply starved — nutritionally, intellectually, emotionally — and that these deficits left them brain damaged. Medical experts and liberal policy makers concurred that the ghetto was having an unquestionably detrimental effect on the learning capacities of poor children of color. Raz observes that a "medicalized, deprivation-based interpretation of low scholastic achievement" therefore came to bolster much of the rhetoric that supported federally funded compensatory educational programming (and most specifically the Head Start program established in 1965) aimed at enriching the lives of inner-city youth.[36]

The diagnosis of minimal brain dysfunction came to be interpreted as entirely distinct from that of mild mental retardation. MBD addressed an elusive neurological condition said to afflict another sociocultural demographic
altogether: the middle-class and white children of the American suburbs. In
these well-to-do surroundings, children could scarcely be said to be under-
stimulated or starved for emotional connection or cultural nourishment.
Instead, these children were perceived to be *over*stimulated and, as a consequence, prone to excitability, restlessness, and impulsivity. They were said
to suffer tantrums and outbursts. They were disorganized in their thinking
and in their memory. Their attention spans were short, and their capacity
to make decisions when confronted with *too many* choices was compromised. Their manner could be described as hyperactive or hyperkinetic, and
their clinical diagnosis might be cerebral dysfunction, hyperkinetic impulse
disorder, hyperexcitability syndrome, or hyperkinetic behavior syndrome.[37]
But these categories now came to be subsumed under the more general um-
brella terminology of minimal brain dysfunction, which, it was increasingly
argued, required psychopharmacological intervention. Terms like "hyperkinetic syndrome" or "organic behavior disorder" did not suddenly vanish;
in 1966 Clements had identified more than three dozen terms for the condition he sought more universally to be known as minimal brain dysfunction. But a move toward consolidation did begin to take place, even as race
continued to have a profound impact on diagnosis and treatment. Whereas
disadvantaged children said to suffer from mild mental retardation received
treatment that entailed enrichment programs, the recommended therapy
for white and affluent children diagnosed with MBD involved prescriptions
of stimulant drugs.

Although these classifications were incoherent and the differences between them obscure and contradictory, in the wake of the Department of
Health, Education, and Welfare report prepared by Clements in 1966, we
see the beginnings of a campaign that portrayed the hyperactive child who
might benefit from a Ritalin prescription as both white and middle-class. An
early illustration of this can be found in the career of Dr. Mark A. Stewart,
a leading researcher into, and an early proponent of, using stimulant drugs
to treat (what he called) "the hyperactive child syndrome." Stewart made
clear in his published accounts that his subjects all came from middle-class
homes.[38] Stewart said nothing about these children's racial backgrounds, but
a profile of his St. Louis pediatric practice (published in 1967) included several photographs of Stewart in his office with a patient—who happened to

FIGURE 4. Psychiatrist Mark A. Stewart in 1967 with a patient diagnosed with hyperactivity and the patient's mother in his clinic at the Children's Hospital in St. Louis. By this time, Stewart had been treating hyperactive children for four years. He was a strong proponent of using stimulant medications to treat hyperactivity, but he also believed in the importance of a stable home environment. In a 1967 profile, Stewart said, "A hyperactive child, say at eight years old[,] is doing more than his share of lying, or impulsive acts, such as taking change from his mother's purse. The child simply acts impulsively, never stopping to think about what he's doing. Things become serious if he is inadequately supervised at home. Each time he gets away with these acts, his tendency for this behavior is reinforced. When he reaches twelve, the physiological problem behind his restlessness may have disappeared; but by then, he may be a budding delinquent. The child is vulnerable in a chaotic home situation. Children from more favorable environments tend not to become delinquent." Reprinted by permission, Washington University Photographic Services Collection, Washington University Libraries, Department of Special Collections.

be a white boy — and his mother (see figure 4). The other photographs along-side this profile were also of white boys. The profile underscored how Stewart planned to treat these children: "About half of all hyperactive children respond dramatically to one of a family of stimulant drugs. This response suggests strongly that the condition has a physiological basis."[39] (Here we also see Stewart's logic at work, for it was not that a measurable organic source of hyperactivity had been located in these children. The diagnosis was based on observation of behavior, but Stewart announced that he calculated that there *had* to be an organic source to these children's difficulties on the basis that the boys appeared to do better once they had been prescribed Ritalin.)[40] Along related lines, a study conducted in Baltimore County, Maryland, in the early 1970s (and published in the *New England Journal of Medicine*) de-termined that far more children attending parochial elementary schools "in the relatively affluent half of the county" were receiving methylphenidate for hyperactivity than were "their less financially fortunate schoolmates" in the poorer half of the county.[41] Here socioeconomic status came to stand in for racial background, even as race was left unmentioned.

The Nature of Nurture

Analyses of race and class were handled more directly in discussions of mini-mal brain dysfunction within the field of psychology. By the end of the 1960s, several leading psychologists began to argue that brain disorders were being drastically underdiagnosed among children in poor and nonwhite commu-nities, despite the fact that such communities likely had disproportionately higher numbers of children suffering from conditions like minimal brain dysfunction and hyperkinesis. In a sense, these psychologists read the de-scription offered by Sam Clements of MBD against itself. Clements had sought to differentiate between the "cultural deprivations" of the inner city and the "minimal brain dysfunction" of more affluent children. These psy-chologists, however, sought to distinguish between those ghetto children whose problems were predominantly environmental and those minority children whose disorders were physiological in nature and who were there-fore less likely to benefit as much as was hoped from such compensatory educational programs — or who would benefit more from compensatory en-richment if they also received medication.

At a 1967 conference on psychological factors in poverty, the influential developmental psychologist Urie Bronfenbrenner spoke about what he saw as the problems inherent in placing an African American child into a racially integrated classroom. Bronfenbrenner said that this new integrated setting might assist that black child "under certain circumstances" but that it was just as possible that "on his arrival there he brings with him his full array of defects and disruptive behaviors." Integration was not going to be enough, Bronfenbrenner observed, unless and until educators and psychologists began fully to acknowledge also the "organic bases of inadequacy" and the "neurological damage resulting in impaired intellectual function and behavioral disturbances, including hyperactivity, distractibility, and low attention span," in so many African American children.[42] In 1969, an article published in the *Journal of Learning Disabilities*, coauthored by clinical psychologist Robert L. Ganter, announced the startling statistic that "over 50%" of "disadvantaged children" in its study had demonstrated neurological deficits (on the basis of interviews and observations and not on the results of electroencephalograms — which typically were "completely normal"). This study argued that "too little attention has been focused on the specific learning disabilities of individual children who live in deprived neighborhoods." It went on to state that it was "simply not adequate to label the poor achiever in the inner-city classroom as culturally deprived and allow this all-inclusive term to explain his poor approach and response to the learning experience." The report added that "the readily apparent socio-cultural inadequacies of the urban ghetto" served merely to mask the "equally significant and alarming . . . major role of neurologically based disabilities" among inner-city children. It concluded, "While neurologic disorders exist to a significant degree in all childhood populations with learning problems, the presence of highly visible socioeconomic handicaps to proper childhood learning abilities easily obscures the perception of the educator or health consultant."[43] Psychologist Edith H. Grotberg also wrote in 1970 in the *Journal of Learning Disabilities* that "the neurological elements" of learning disabilities tended too often to be "overlooked or buried in the generally poor test performance of impoverished children," a fact that must be understood as "indefensible in the light of research evidence."[44] The worry of these researchers was not that schools were singling out poor and nonwhite children as suffering from neurological deficits; the concern was that these children's deficits in brain function were most likely being misrecognized and underdiagnosed *because* they were black and poor.

Some educational psychologists soon began to argue directly that also "disadvantaged" children of color deserved to be entered into the ranks of those diagnosed with MBD (especially when their symptoms were evidently no different from those of the white children who were being diagnosed). The *Journal of Learning Disabilities* emerged as an important venue where psychologists could voice their critique that minority students were being drastically underrepresented in the count of children diagnosed with minimal brain dysfunction. Educational psychologist Daniel P. Hallahan, an authority on how to teach children with learning disabilities, speculated in early 1970 that "it may be that, because of a deprived or disordered early life, the lower class child is likely to have cerebral dysfunctioning of some kind."[45] Psychologist Lester Tarnopol, also an expert on special education, surmised in the journal's pages that spring how the "cumulative evidence tends to support the hypothesis that a significant degree of minimal brain dysfunction exists in the minority group, delinquent, school dropout population."[46] Wishing to underscore these points, the journal had a full-page photomontage of schoolchildren accompany Tarnopol's article—photographs that portrayed only African American children (see figure 5). Using black children to illustrate an article on the "hyperkinetic, brain-injured child" was presumably meant to advance a (not-so-subtle) position that schools were insufficiently attuned to the multiplicity of ways "deprivations" afflicted poor African American children.[47] Social factors did not always explain these children's educational needs.

In this way, psychologists worked to shift a perception of who was most at risk of MBD—a strategy that aimed quite literally to change the face (and race) of the brain-injured child in the United States. The argument was that the incidence of minimal brain dysfunction was likely much higher among children in black and poor communities than was being diagnosed. There was the additional view that poor and black children probably suffered from subclinical brain injuries far more often than white children and that their educational needs were being neglected in the process.

Yet an unintended consequence of such claims was an incapacity to reconcile the confounding relationship between environmental and biological factors. Did their disadvantaged environment result in more brain damage among African American youth? Or were black children more prone to MBD due to other—possibly hereditary—factors? These questions made some psychologists quite uneasy. Some experts, sensitive to the long-standing denigration of black children's intellectual capacities, fretted that any entry into the category could inadvertently exacerbate the pathologization of these children. A report from a team of clinical psychologists in 1970 that addressed

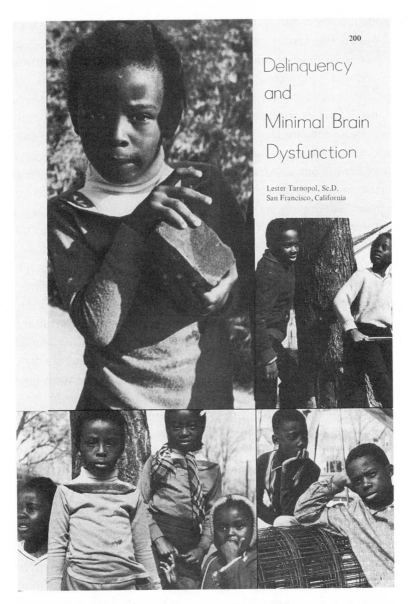

Delinquency
and
Minimal Brain
Dysfunction

Lester Tarnopol, Sc.D.
San Francisco, California

FIGURE 5. This photomontage accompanied an article by psychologist Lester Tarnopol, "Delinquency and Minimal Brain Dysfunction," in the April 1970 issue of the *Journal of Learning Disabilities*. Here, Tarnopol speculated that MBD might be far more widespread among minority children than generally assumed—and that this "may partially explain why the special programs to help educate this population have tended to lack success."

the epidemiological distribution across socioeconomic and racial divisions of brain dysfunction starkly concluded that "from all indications, CNS [central nervous system] dysfunction is not randomly distributed in the population at large" and that "cases of brain damage in children are highly concentrated in the poor white and black segments of the community"—although the problem "appears to be considerably more extreme in the case of black children of lower socioeconomic origin relative to their white counterparts." However, the authors of this report acknowledged that they were "extremely ambivalent concerning our data," adding—with what might be characterized as understatement—that their findings were "potentially both destructive and dangerous." Attempting to navigate the narrow straits between their own assumption that "*race* itself is largely a biological concept" (while "*racism* is a sociological concept") and their desire to find social explanations for the disproportionate predominance of brain dysfunction among children of color, these psychological researchers emphasized that "underlying social problems" were always going to be far more influential than biological factors. Their context-sensitive hypotheses, however, were not ultimately supported by conclusive empirical evidence, as they ended up finding (in a study in Muskegon, Michigan) more brain damage among middle-class white children than their research design had led them to expect but also extraordinary rates of brain dysfunction among poor children of color. "The prevalence of brain damage appears to be extremely high in certain sectors of the population—particularly in the black subculture—and in those blighted regions of the community where poverty is a way of life," they stated.[48] The combination of discomfort at countenancing a biological explanation and difficulty in proposing meaningful alternatives to the role "racial" factors might be playing left psychologists at an impasse that emerged directly out of an avowedly liberal and anti-racist stance.

It was an impasse that Berkeley psychologist Arthur R. Jensen (to choose a prominent example) was perfectly willing to exploit for his own ends. In his book *Educability and Group Differences*, Jensen cited extensively from this same 1970 epidemiological report on the unequal racial distribution of MBD in the population. He additionally acknowledged Urie Bronfenbrenner's pronouncements on the neurological disorders he believed to be disproportionately evident among poor and black children. Jensen summarily dismissed what he considered to be an "ideologically motivated insistence that all such effects must be attributable entirely to external environmental conditions." Impoverishment could not explain everything, Jensen insisted, because poor whites were not suffering as dramatically as poor blacks from neurological deficits. "Something more seems to be involved than just socioeconomic conditions," Jensen coyly observed before moving to his barbed perspective that differences in rates of neurological deficits between whites and blacks could in fact turn out to be explained with a detailed investigation of "the ancestral gene pools of the American Negro." Not that Jensen necessarily had proof for "a largely genetic explanation of the evidence on racial and social group differences in educational performance."[49] But as a review in the *American Journal of Psychology* shorthanded it, Jensen's prose on race and education was rather "akin to screams of 'fire' in a crowded theater": it was guaranteed to provoke strong reactions.[50]

The Speed of White

After the sensational reports from Omaha made national headlines, the debate over stimulant drug treatments for schoolchildren rapidly polarized. On the one side, there remained the unchecked enthusiasm of child psychiatrist Eric Denhoff, who—soon after the Omaha story—still chose to inform the *New York Times* that Ritalin remained likely in the long run to be seen as "the penicillin of children with learning disabilities."[51] Psychiatrist Maurice Laufer, Denhoff's colleague, voiced a comparable conviction that stimulant medications remained an excellent means to treat hyperactivity, as did psychiatrist Paul Wender, who never wavered from his view that "the most effective single treatment for MBD children is medication."[52] Ciba, the Swiss pharmaceutical manufacturer of Ritalin, ran numerous advertisements in the course of the 1970s (all profiling white boys), some of which self-reflexively took up the criticism that MBD might be a bogus medical category. These ads went on to rebut the criticism with assurances that there was now "psychometric testing" available to establish the diagnosis of MBD and that Ritalin was indeed the best way to remediate the condition (see figure 6).

On the other side, however, there was the furious rhetoric of commentators like journalist Nicholas von Hoffman, who declared that while dissidents in the Soviet Union were being diagnosed "as crazy" and put "in the 'loony bin,'" in the United States "we say of a kid who won't go along with some soporific program that he's got marginal brain damage and we dope him up." To give a stimulant drug to a child as "a therapeutic procedure," von Hoffman added, was nothing but a "biochemical" means of putting "the cop

FIGURE 6. This advertisement for Ritalin appeared in the *American Journal of Diseases of Children* in June 1971. Strikingly, the text self-reflexively takes up the ambiguity of whether MBD was real or invented, as it instructs pediatricians to take care to keep up with the latest medical breakthroughs. "What medical practitioner has not, at one time or another, been called upon to examine an impulsive, excitable hyperkinetic child?" the text rhetorically asks. It continues: "In the absence of any detectable organic pathology, the conduct of such children was, until a few short years ago, usually dismissed as 'a phase,' spunkiness, or evidence of youthful vitality. But it is now evident that in many of these children the hyperkinetic reaction syndrome exists as a distinct medical entity." Every advertisement that Ciba, Ritalin's manufacturer, ran at this time portrayed the children afflicted with MBD as white.

and the concentration camp inside the pill" so that we can "put the pill inside the kid."[53] Psychiatrist Lester Grinspoon was only somewhat more temperate when he remarked that "the use of drugs, particularly amphetamines, in the name of therapy often does little more than provide a relatively easy and economical way of making the classroom situation more tolerable and manageable for the teacher."[54] And there was sociologist Peter Conrad in his book *Identifying Hyperactive Children*, declaring that — with a little help from psychopharmaceuticals like Ritalin — modern medicine stood poised to function "as an agent of social control."[55]

Popular accounts of stimulant drug treatments for children were far more sympathetic. Nearly every leading women's magazine — including *McCall's*, *Redbook*, *Woman's Day*, *Parents Magazine*, and *Ladies' Home Journal*—ran positive articles on the topic, sometimes written by major authorities on children's health like pediatrician T. Berry Brazelton and psychologist Bruno Bettelheim.[56] These articles frequently arrived with distressing titles (like "The Children Who Can't Sit Still," "When a Child Has Trouble Learning," and "The Child Who Wasn't Retarded") that served to alert middle-class mothers to the prevalence of hyperkinesis and the potential benefits of stimulant medications like Ritalin.[57] Book-length manuals informing concerned parents on the early warning symptoms of hyperactivity — as well as advocating for stimulant drug treatments — also began to reach a broad mainstream audience. Here too a central intent was to assuage parents' fears about putting their children on stimulant medications. Mark A. Stewart's *Raising a Hyperactive Child*, for instance, movingly cited the testimony of a seven-year-old who informed his mother, "I have friends when I'm on Ritalin."[58]

While these writings aimed at middle-class parents declared Ritalin the avenue of first resort for hyperactive (white) children, a cloud of concern about racial discrimination continued to hang heavily over this same choice when it came to minority children. There remained widespread anxiety and anger that stimulant drugs were being routinely *overprescribed* to African American children. After the *Christian Science Monitor* in late 1970 reported that "medical experts" stated that hyperactivity likely afflicted 30 percent of children "in ghetto areas," the Rhode Island chapter of the Students for a Democratic Society in 1971 distributed a flyer pithily titled "Fight Racist Drugging."[59] The liberal social science journal *Trans-action* echoed these same statistics in "Drugging and Schooling," an essay that announced how "at least 30 percent of ghetto children are candidates" for "amphetamine and stimulant therapy."[60] The *New Republic* observed the same year, "The advent of the 1971–1972 school year means a return to the Ritalin regimen for thousands of school children, a great many of them from poor families in ghetto schools."[61] Numerous stories circulated about teachers targeting their minority students for stimulant drug treatments. A journalist in New York reported on a white elementary school teacher who did not behave toward her Latino pupils "as nicely as she did the others, becomes angry and screams often, throws books, and uses words like 'moron' and 'idiot' when a child does not do well on a particular lesson." This teacher recommended that her Latino students get medical exams — whereupon they received prescriptions for Ritalin.[62] In 1972, an African American community group based in Brooklyn additionally announced its plan to file a class-action lawsuit against the New York Board of Education "to halt the administration of the drug, Ritalin, in the predominantly Black elementary schools."[63] Further, reports arose during the 1974–75 school year that 80 percent of the children selected to participate in a stimulant drug program in Minneapolis were African American.[64] There was no trust and no faith among many in the black community that stimulant medications were not simply a ploy to subdue the allegedly disruptive behavior of the inner-city child of color.

Leading African American intellectuals echoed communal doubts when they spoke out loudly against stimulant drug treatments. William B. Banks, chair of the Afro-American Studies Program at the University of California, Berkeley, observed, "A school system that is short on funds and long on active and aggressive black youngsters might very well bypass the necessary neurological and psychological examinations to determine hyperkinesis." Banks pointedly added that "drug treatments for children represent yet another example of a dominant cultural group imposing its own terms and limitations on other groups."[65] Marian Wright Edelman, director of the Children's Defense Fund, said in 1974 — echoing sentiments expressed a few years earlier by Leon Eisenberg — that "if a child is restless because he's hungry, adequate nutrition ought to be the first level of inquiry and response — not the administration of some drug."[66] After he had helped to shut down a project in the Boston public schools that planned to test the effects of psychotropic drugs (including Ritalin) on children diagnosed with minimal brain dysfunction — and that local groups asserted would likely be used to target black youth — community activist Melvin H. King said in 1976, "People might say that I'm paranoid about my relationship as a black person to white people of power in this society, but I don't think so. I think we've seen enough of what people have attempted to do to people whom they perceive to be different and powerless."[67] And in 1978 Cornell psychologist A. Wade Boykin wrote about "the so-called hyperkinetic syndrome" that it might simply be the case that African American children possessed "a rich movement repertoire" and that therefore schools provided them "inadequate affordance of stimulus change and inadequate allowance for the expression of behavioral variability."[68] In this highly charged and divisive climate, it became virtually impossible for psychologists — or anyone else for that matter —

to urge that *more* stimulant drug treatments be made available to inner-city African American children. And no one really did.

What *could* still be said, however, was that the racial background and class status of a child did make a profound difference in terms of how a child's learning difficulties were classified. A study supported by the National Institute of Mental Health in 1971 found conclusive evidence from several epidemiological surveys that "disproportionately large numbers of persons from low socioeconomic status and from ethnic minority backgrounds [are] among those persons labeled as mentally retarded" and that — at least in California, where the surveys were conducted —"rates of placement for Mexican-American and Negro children in special education classes . . . were two to three times higher per 1,000 than rates for children from English-speaking, Caucasian homes."[69] At the same time, two psychologists in 1976 observed that "there is a higher proportion of children diagnosed as learning disabled from families of higher socioeconomic status than there is from families of lower socioeconomic status" and that this was likely due to the fact "that middle-class parents have higher education aspirations for their children, and are better able and willing to articulate their concern when their aspirations are not fulfilled." Additionally, they noted that such stark differentials in diagnoses across socioeconomic lines could well be the result of "the prejudgment" among "diagnostic personnel" that "low achievers in inner city schools are common retardates while their cousins in the suburbs are 'learning disabled.'"[70] Although "the minority child" diagnosed as "educable mentally retarded" bore "striking similarities" to the white and middle-class child diagnosed with a learning disability, journalist Peter Schrag and educator Diane Divoky observed acutely in *The Myth of the Hyperactive Child and Other Means of Child Control* in 1975 that this "way of disassociating middle-class children from blacks and other minorities" served only "to maintain belief in the dynamics of class success, the belief that white, affluent, well-mannered parents simply cannot produce offspring who can't behave and can't learn and are therefore no better than ghetto blacks."[71] In 1979, the *Journal of Learning Disabilities* summarized what amounted to a growing consensus among psychologists and educators that "children from the lower socioeconomic class have increased possibilities of being identified as retarded, while children from the higher socioeconomic classes are more likely to be the recipients of the learning disabled tag."[72] By 1980, an educational psychologist wrote in the journal *Learning Disability Quarterly* that "the present situation finds the [poor] child doubly penalized, first because of cultural background or socioeconomic status, and secondly, because of deficit labels which tend to be more stigmatizing for the child when referred for special education service." It amounted, this author observed, to a grotesque form of "'special miseducation.'"[73]

What also remained possible to say in this polarized climate was that the estimated number of all hyperactive children receiving stimulant drugs continued to climb — though precise numbers proved extraordinarily difficult to pin down. In 1970, a congressional hearing "on the use of behavior modification drugs on grammar school children" suggested a range of between 150,000 to 200,000 schoolchildren receiving stimulant drugs.[74] When a survey in the Washington, D.C., area asked physicians in 1971 to report whether they had prescribed stimulant drugs to hyperactive children, more than 90 percent of these doctors reported that they had done so.[75] When researchers in Baltimore County followed up on a study in 1971 with a second survey in 1973, they found an increase of 62 percent in schoolchildren who were receiving drugs — and that the majority of these prescribed drugs were stimulants.[76] News reports extrapolated local results like these to the national level. In 1973 *Time* magazine estimated the number of schoolchildren who were taking stimulant drugs as treatment for hyperkinesis to be approximately 300,000; *Time* reported the anecdotal story of a teacher in California who had "recommended drug therapy for nine of her 28 pupils because their spirited behavior convinced her that they were brain-damaged."[77] In 1975 Schrag and Divoky used data collected by the National Disease and Therapeutic Index and the National Prescription Audit, both based on statistical samplings of physician records (and used by pharmaceutical manufacturers to gather marketing information), to conclude that between 600,000 and 700,000 prescriptions for stimulant drugs were being written for schoolchildren to treat hyperactivity and minimal brain dysfunction. They claimed that the numbers of children taking stimulant drugs "have been doubling every two or three years."[78]

There were even more disparities in estimates of schoolchildren who might benefit from stimulant drug treatments. In 1969, before news of the Omaha experiment broke, a clinical psychologist and a mental health consultant in the public schools in upstate New York surmised "that 30 to 50% of children referred to us because of disruptive classroom behavior and/or poor academic achievement" could be helped more from stimulant drugs than from special educational programming or psychotherapy.[79] In the fall of 1970, after the news from Omaha went viral, the *Christian Science Monitor* reported (based on estimates from the National Institute of Mental Health) that "there are up to 4 million 'hyperactive' children in the United States" and that every one of these children might likely benefit from drugs like Dexedrine and Ritalin.[80] The purported number of children with minimal brain dysfunction in the United States ranged even higher. By the mid-1970s, there were claims that "in clinics treating psychiatrically disturbed, primary-school age children, MBD, broadly construed, probably constitutes at least 50% of the cases referred."[81] There were even reported estimates that as many as 15 to 20 percent *of all schoolchildren* in the United States suffered from minimal brain dysfunction or hyperkinesis.[82]

Such staggering statistics on the prevalence of hyperactivity and minimal brain dysfunction in American children left observers both incredulous and dismissive. As Lester Grinspoon wrote, "It is impossible to believe that the 200,000 or more school children who are now being routinely administered stimulants are all suffering from organic brain damage or deficiencies in crucial CNS chemicals."[83] Others vaunted credible alternatives to medications, notably pediatric allergist Ben F. Feingold, who, in his best-selling book, *Why Your Child Is Hyperactive*, proposed that a food "elimination diet" (such as the complete removal of artificial food colors and flavors from a child's meals) would be far more effective in the treatment of hyperactivity in children than the "doubtful therapy" of stimulant drugs.[84] Yet even here the issues of race and class were a scarcely concealed subtext. Insisting that parents of hyperactive children purchase only unprocessed and natural foods, the Feingold diet was so costly that affluent (and often also white) families were the only ones likely able to afford it.

Conclusion

The era when the nonspecific and largely speculative term "minimal brain dysfunction" would be used to describe the hyperactive child came to an end in the course of the 1980s. By this time, attention deficits in children — with or without the presence of hyperactivity — became the primary focus for many psychological researchers; attention deficits, like hyperactivity, had been demonstrated already in the 1970s as also treatable with stimulant drugs.[85] This new focus on cognitive inattention — rather than on hyperkinetic movement — came finally to transform the disorder's conceptualization, a shift codified in 1980 in the third *Diagnostic and Statistical Manual of*

Mental Disorders (*DSM-III*), which introduced new classifications of attention deficit disorder (ADD) with hyperactivity (ADD/H) or without it (ADD/WO). This terminology would be modified in 1987 in the revised edition of the *DSM-III*; here attention-deficit/hyperactivity disorder (ADHD) became a single entity—only to be subdivided once more in *DSM-IV* (1994) into three subcategories: hyperactive/impulsive (ADHD/HI), inattentive (ADHD/I), or both hyperactive/impulsive and inattentive (ADHD/C). *DSM-5* (2013) left the criteria for the ADHD diagnosis largely unchanged.[86]

In the course of these same decades, diagnoses of ADHD and prescriptions for stimulant drugs both continued to skyrocket. By 1993, more than 1.3 million children were being prescribed Ritalin for ADHD.[87] By 2009, the Centers for Disease Control estimated that 9 percent of all children in the United States had ADHD.[88] ADHD began additionally to be diagnosed into adulthood. Controversy over the validity of the ADHD diagnosis—and the use of Ritalin to treat it—also still raged.[89] The skeptical view remained that the disorder was nothing other than a means "to label difficult children who are not ill but whose behavior is at the extreme end of the normal range."[90] Methylphenidate use continued to expand dramatically as well. From 1981 to 1987, its usage doubled—and doubled again from 1990 to 1995.[91] As a critic wrote in 1996, children perfectly capable of sitting "quietly and [who] perform well in social situations or in one-on-one psychometric testing can still be candidates for the diagnosis and treatment of ADHD"—and hence candidates for stimulant drugs—"if their parents or teachers report poor performance in completing tasks at school or at home."[92] By 2013 the *New York Times* reported that one in seven children in the United States received an ADHD diagnosis—and that 3.5 million children were being prescribed stimulant medications to treat the condition. This epidemic of ADHD diagnoses prompted psychologist Keith Conners, who had participated in the earliest drug trials with hyperactive children half a century earlier, to declare the situation nothing less than "a national disaster of dangerous proportions."[93]

Race and socioeconomic status remained an often unstated presence in these debates. Reports surfaced in the first years of the twenty-first century that Ritalin was now being routinely taken off-label for performance enhancement. The President's Council on Bioethics put the problem like this: "Do we want a society where a non-medicated person can't compete successfully?"[94] There were by this time frank discussions that a diagnosis of ADD or ADHD amounted to a "loophole" that guaranteed the children of the well-to-do an extra edge when they applied to college. *Forbes* magazine in 1996 observed that affluent parents actively pursued a diagnosis for their children so that they would have an untimed Scholastic Assessment Test (SAT) score administered in a private (distraction-free) setting. "ADD has become an enormous and growing business," an Ivy League admissions officer candidly told the magazine. "It's also a rich kid's business and an enormous class issue." The same applied to a Ritalin prescription — since "getting a doctor's certification" for the drug was often "an expensive process."[95] In 1997 the *New Republic* magazine ran a satiric cartoon of a white infant with a silver spoon in his mouth with a caption that read, "What Does Your Healthy, Normal, Perfect, Little Darling Need to Get Ahead in Life? A Small Disability to Qualify for Special Aid!" (See figure 7.) The accompanying article spelled out a similar theme: "Sometimes, it seems, the problem is less inattentive children than overattentive parents, many of whom are unwilling to believe their progeny is less than perfect."[96] The Heritage Foundation unabashedly spun the news more specifically to stir working-class resentments, asserting in 1999 (in *Policy Review*, the right-wing think tank's flagship publication) how the diagnosis of ADHD "can operate, in effect, as affirmative action for affluent white people."[97]

Analyses like these unhelpfully ignored the open secrets surrounding the ADHD of African American children. In 1997, and despite the astronomical rise in rates of ADHD diagnoses in the United States, a team led by an African American clinical psychologist noted that there remained an "unexplored void of ADHD and African-American research" as well as an "extraordinarily limited literature on the pharmacotherapy of ADHD in African-American children."[98] In 2009, a follow-up report determined that "African American youth are diagnosed with ADHD at only two thirds the rate of Caucasian children," even though "in light of our review of the evidence, it seems unlikely that African Americans are exhibiting fewer symptoms." This evidence prompted the report to conclude that "there is an urgent need for better education of the African American community about ADHD, as well as development of treatment approaches that are perceived as acceptable to parents"— including possible alternatives to (what many in the black community perceived to be the potentially addictive qualities of) stimulant medication therapies.[99]

When psychologist Sam Clements first advocated for a concept of minimal brain dysfunction in the early 1960s, his aim had been to destigmatize learning disabilities of white children. The term in its essence was always inescapably racialized. Conflating race and class, Clements took the

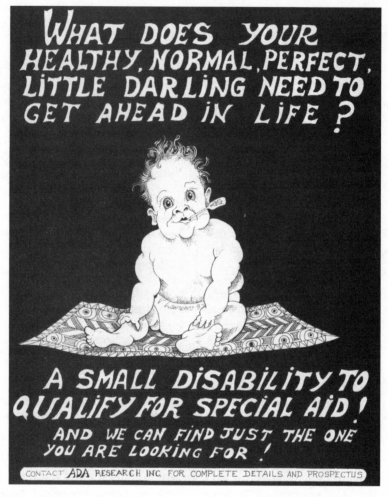

FIGURE 7. This cartoon by Vint Lawrence appeared in the *New Republic* on August 25, 1997. The accompanying article by journalist Ruth Shalit documented the ballooning use of diagnoses of "learning disability"—including Attention Deficit Disorder but also such creative, newly fabricated "hidden impairment[s]" as "dyscalculia" (a "math disability") or difficulties with "phonological processing"—that were proving most helpful for otherwise "fast-track students" to achieve significant accommodations in "hyper-competitive" university settings. Courtesy of Anne Garrels.

symptoms associated with MBD — most especially hyperactivity — to be a condition experienced principally by white and middle-class children. Black children who were poor required cultural enrichment programs like Head Start, while white or middle-class children needed stimulant drugs (and a nonpathologizing diagnostic category). Meanwhile, the fact that African American parents were so resistant in the course of the 1960s and 1970s to medication treatments for their children can hardly be surprising in the context of the era's often brutal and mendacious racial politics. That the racial politics of stimulant drug use for children diagnosed with ADHD remains both confused and contradictory half a century later may well be an inadvertent outcome of this tangled and troubled history.

THREE

The Politics of Cerebral Asymmetry
and Racial Difference

I N THE COURSE of the 1970s and 1980s, a development within the rarefied field of neuropsychology swept unexpectedly into the popular American mainstream. This was a widespread fascination with split-brain research, sometimes known as "dichotomania," and it extrapolated from experimental findings regarding a couple of neuroanatomical facts about the mammalian brain — namely, that it is divided into a right and a left hemisphere and that there are specialized functions in each cerebral hemisphere — only then to make a remarkable range of claims about what the two hemispheres can (or cannot) do.[1] It became customary in the course of these decades to voice a desire to access one's own right-brain intuitive talents or to state openly that one's rational left brain was simply too dominant and needed to be mellowed — and that it was possible (with practice) to do so. Influential psychologists as well as many leading public intellectuals in a variety of disciplines across the humanities and hard sciences contributed to these discussions in well-regarded general interest magazines and journals like *Scientific American, Saturday Review,* and the *New York Times Magazine.*

In 1976, for instance, iconoclastic management guru Henry Mintzberg effused (in the pages of the *Harvard Business Review*) that split-brain research had "great implications for both the science and art of management." He declared that management theorists too frequently neglected the creative capacities of "well developed right-hemispheric processes," choosing to focus instead on "the logical, linear functions" that resided in the human brain's left hemisphere. "Did you ever wonder why some things come so easily and others seem so difficult, why sometimes you just cannot get your brain to work?" Mintzberg rhetorically inquired. "Maybe the problem is not that you are stupid or tired, but that you are tackling a problem that taxes the least developed hemisphere of your brain."[2] This conception of cerebral asymmetry — and the notion that skills inherent in our right hemisphere went too routinely unacknowledged and underutilized, while left hemispheric processes were likely often overdeveloped — was to gain remarkable traction in the years that followed.

By the early 1980s, a veritable trove of self-help books urged readers to learn to utilize *both* hemispheres of their brain. People should engage in *whole-brain thinking* and not neglect the untapped potentialities of their right brain. Many of these books — published in the wake of economic recessions in the mid-1970s and early 1980s — offered financial advice; these included *The Tao Jones Averages: A Guide to Whole-Brained Investing* (1983), *Whole-Brain Thinking: Working from Both Sides of the Brain to Achieve Peak Job Performance* (1984), and *Intuitive Management: Integrating Left and Right Brain Management Skills* (1984). Marketed as an asset to anyone in pursuit of an extra edge (at work or in life more generally), split-brain theory became a lucrative enterprise. Yet declarations about the virtues of right-hemispheric skills were hardly limited to the question of how to improve one's chances of making it on Wall Street.

Neuropsychological insight into the right and left hemispheres ranged into realms that were far more esoteric, if no less influential. Notably, for instance, Princeton psychologist Julian Jaynes hypothesized in 1976 in *The Origins of Consciousness in the Breakdown of the Bicameral Mind* that schizophrenia could be traced to the asymmetric hemispheres of the brain and that the voices heard by schizophrenics emanated from the right temporal cortex. These voices had become stigmatized in the course of human history and were now largely taken to signal mental illness. However, Jaynes averred, the hallucinatory voices heard by schizophrenics should be interpreted as "the voices of the gods," since it was entirely possible "that there exists [a] vestigial godlike function in the right hemisphere."[3] Jaynes's book was a great critical and commercial success; it received a nomination for a National Book Award in Contemporary Thought (and has remained in print to the present day).[4] There were as well the mind-altering implications set out by Cornell University astrophysicist Carl Sagan in his own best-selling book of 1977, *The Dragons of Eden*: "I wonder if, rather than enhancing anything, the cannabinols (the active ingredients in marijuana) simply suppress the left hemisphere and permit the stars to come out." To which Sagan added, "This may also be the objective of the meditative states of many Oriental religions."[5] And there was art teacher Betty Edwards's blockbuster guide-book of 1979, *Drawing on the Right Side of the Brain: A Course in Enhancing Creativity and Artistic Confidence*, which directly linked cerebral specialization to the capacity for creativity, observing, for instance, that "the emphasis of our culture is so strongly slanted toward rewarding left-brain skills that we are surely losing a very large proportion of the potential ability of the other halves of our children's brains."[6] Edwards's book would (in multiple editions and translations) sell more than two million copies worldwide.

There was another consequential — if more often overlooked — way in which the neuropsychology of cerebral specialization was to have a profound social and cultural impact in the course of the 1970s and 1980s. This had a great deal to do with core controversies within the fields of school and educational psychology about how most effectively to instruct children and how equitably to measure children's intelligence. Sharp disputes emerged after Berkeley psychologist Arthur R. Jensen, the prominent and influential expert on IQ and race, had begun repeatedly and at length to argue that since "SES [socioeconomic status] intellectual differences" were "due largely to genetic factors"— and could not be "entirely attributed to environmental differences"— it could come as no great surprise that "intensive efforts by psychologists, educators, and sociologists to devise 'culture-free' or 'culture-fair' tests . . . have not succeeded."[7] Jensen's position that there was a key genetic basis for human intelligence, and that therefore African American children as a cohort were not capable of achieving IQ results comparable to the intelligence scores of white children, spurred heated debate among educators and psychologists who assailed Jensen's evidence and countered his views that there existed an immutable link between IQ, race, and socioeconomic status.[8]

However, neuropsychologists often took another tack; they tended not to quarrel with Jensen's claim that there were intellectual differences between black and white children. Yet even as they opted tacitly to accept Jensen's evidence, they nonetheless challenged his conclusions. The differences in intelligence between white and nonwhite children, neuropsychologists argued, were either poorly understood or completely misunderstood. It was not, they proposed, that nonwhite children possessed an insurmountable shortfall in their collective IQ, as Jensen insisted, but rather that minority children possessed another source of intelligence altogether: right-hemispheric intuitive skills and creative, nonverbal, and spatial abilities that were not being recognized by extant intelligence measures. In short, a neuropsychological position often declared that minority children — African American, Latino, and Native American, in particular — had talents that white children typically lacked. Rather than being worse off, these nonwhite kids were — in quite a number of important respects — *better off* than white children. That the American school system remained neither willing nor ready to validate these children's right-hemispheric talents, and that IQ tests failed to quantify these right-brain skills, quickly led educators and psychologists passionately to demand reforms in educational policies, IQ testing methodologies, and school curricula more generally.

The Return of Split-Brain Research

Theories of the split brain in humans were not new developments in the 1970s and 1980s. More than a century earlier, French surgeon Paul Broca reported that patients with brain lesions in the left frontal lobe experienced a loss of speech. Consequently, Broca posited that the left hemisphere was the location in the brain for language. Karl Wernicke, a German neurologist, also proposed that the left side of the brain controlled speech and language functions. More research concerning the lateralization of cerebral faculties soon followed. Based on his clinical observations, British neurologist John Hughlings Jackson suggested in the 1860s and 1870s that the process of visual ideation was likely localized in the brain's right hemisphere. During the next several decades and into the first half of the twentieth century, research into hemispheric specialization significantly subsided, although important work was done by the German neurologist Hugo Liepmann to establish the role of the left hemisphere in purposive motor activity and to reconceive the functioning of the corpus callosum (the neural fibers that connect the two cerebral hemispheres).[9]

In 1940 the first split-brain surgery, or cerebral commissurotomy, was performed. This involved the severing of a monkey's corpus callosum.[10] In the course of the 1950s neuropsychologist Roger W. Sperry at the University of Chicago and his graduate student Ronald E. Myers conducted split-brain studies on cats, which confirmed that the corpus callosum served to transfer visual data from one hemisphere to the other.[11] Sperry further observed that when the corpus callosum was severed in animals (such as monkeys and cats), not only did this prevent "the spread of learning and memory from one to the other hemisphere," but also it "was as if each of the separated hemispheres had a complete amnesia for the experience of the other, as if each had its own independent perceiving, learning, and memory systems." Sperry began tentatively to propose that the two hemispheres of the mammalian brain were "special and nonsymmetrical."[12] At another moment, Sperry declared even more emphatically, "In these respects it is as if the animals had two separate brains."[13] As a leading physiologist acknowledged in 1961, with reference to the discoveries being made by Sperry concerning the asymmetrical roles played by the brain's two hemispheres, the nation's neuroscientists were now confronted with a metaphysical question it would not have seemed possible to ask even a few years earlier: "Why do we have two brains?"[14]

It was in the early 1960s as well that Sperry, now based at the California Institute of Technology, and with the assistance of psychologist Michael S. Gazzaniga and under the supervision of neurosurgeons Joseph E. Bogen and Philip J. Vogel, conducted research with a human subject who suffered from epilepsy and who had consented to a commissurotomy. Bogen proposed that cutting the patient's corpus callosum might prevent the epilepsy from spreading between the brain's hemispheres. The surgery worked; it halted the patient's seizure activity. The patient afterward told Sperry that he felt "better generally than he [had] in many years." But the brain surgery also resulted in "impairments," such as an inability to "cross-locate with either hand across to the other." There was a similar difficulty with regard to visual perception. Sperry concluded that now in a human patient there existed compelling evidence that "the separated hemispheres were each unaware of activity going on in the other in the case of those functions that are highly lateralized."[15]

Sperry and his team subsequently went through a several-year phase in which they mused about the various possible implications of the fact that humans did not so much have a split brain as a doubled one. They continued to mull over how the two halves differed. In 1964, Sperry published the results of his experiments on the bisected brain in *Scientific American*. Here he presented the hypothesis that the "two halves of the mammalian brain are mirror twins, each with a full set of centers for the sensory and motor activities of the body," and that when the connection between these halves was severed, "either half of the brain can to a large extent serve as a whole brain"[16] (see figure 8). Meanwhile Gazzaniga, with rather contrary emphases, affirmed that in adults there was a high level of "hemispheric inequality" and that the right hemisphere handled nonverbal cues while the left hemisphere managed language abilities.[17] And Jerre Levy, another psychologist who studied with Sperry at Caltech, argued that there might be evolutionary

FIGURE 8. Neuropsychologist Roger W. Sperry at the California
Institute of Technology around 1970. Sperry expressed repeated
scorn at what he saw as the bias that many educators had against
students' right-hemisphere skills. For instance, Sperry critically
observed in 1975, "In our present school system, the attention given
to the minor hemisphere of the brain is minimal compared with
the training lavished on the left, or major, hemisphere." In 1981,
Sperry would win the Nobel Prize "for his discoveries concerning
the functional specialization of the cerebral hemispheres." Reprinted by permission, courtesy of the Archives, California Institute of Technology.

reasons for the "differential perceptual capacities" of the right and left hemispheres of the brain.[18] Levy additionally asserted that the female brain might
be less lateralized than the male brain, a fact that she said probably inter-
fered with women's spatial abilities and likely had an "evolutionary basis."[19]
Psychologists in the United States and in Canada continued to pursue and
promote Levy's theory about "sex and the single hemisphere" throughout
the 1970s.[20]

However, it was the neurosurgeon Joseph Bogen who did more than anyone to articulate and popularize concepts specifically about the brain's right
hemisphere. Bogen's ideas meshed well with the countercultural mood of
the historic moment. Already in 1970, for instance, Bogen speculated that
"the psychedelic effects reported by users of LSD may depend upon an increased access to the left (speaking) hemisphere of visuospatial activity going
on in the right hemisphere but ordinarily less available to the left hemisphere consciousness."[21] More significantly, between the years 1969 and 1972
Bogen published a multipart essay titled "The Other Side of the Brain" that
no doubt gained notoriety not least due to its author's stature as a physician
on Sperry's team. "What is the right hemisphere for, in the human scheme
of things?" became Bogen's overarching question. At least at first, Bogen
remained tentative: "We do not yet understand *how* the one hemisphere produces language; but of the other hemisphere we do not even know *what* it is
producing. . . . The lesser known and hence more fruitful area for investigation of mental activity is that carried out by the other side of the brain."[22]
Bogen went on to claim that the right hemisphere was the seat of musical
ability as well as artistic and creative talents (see figure 9).

Bogen additionally pushed his speculations into even woollier terrain.
He noted that persons who lived in "the East" and tribal peoples around
the globe intuitively grasped the inherent duality of human thought (while
Westerners—in his view naively—persisted in presuming a single, unitary mind). Bogen wrote that human beings had accepted for centuries that
the world was flat but that "the strength of this conviction" had been "no
assurance of its truth"; they had also (wrongly) believed in "spontaneous
generation" and "the inheritance of acquired characteristics." Likewise, that
an "inner conviction of Oneness is a most cherished opinion of Western
Man" did not mean that this belief was not misguided.[23] Bogen quoted art-
ists and intellectuals including (but not limited to) psychiatrist Carl Jung,
sculptor Henry Moore, painter Joan Miró, and poet Paul Valéry who had
also understood the profound limitations of an over-focus on rationality.
Jung, for instance, had opined that "every creative person is a duality or a
synthesis of contradictory aptitudes." On this basis, Bogen argued that individuals possessed of a "technical proficiency in music, drawing or writing"
nonetheless often produced art "devoid" of genuine intuitive talent because
"creativity requires more than technical skills and logical thought." He hypothesized further that "certain kinds of left hemisphere activity may directly suppress certain kinds of right hemisphere action" and that there may

FIGURE 9. Neurosurgeon Joseph E. Bogen in a photograph taken in Switzerland in the mid-1980s. By this time, Bogen's theories had begun to have a major impact on efforts by educators to reform intelligence testing techniques so as to measure with more thoroughness skills associated with both sides of the brain. Courtesy of Meriel Bogen Stern.

be "a physiological basis" for the "failure" of some people "to develop fresh insights (in the sense of new understanding of the outside world)." Thus, and according to Bogen, the left hemisphere exerted an "inhibitory effect" on intuition and creative processes.[24] And he additionally, and as it happens quite influentially, began to venture that "reliance upon 'time' as a principle of organization may better distinguish the left from the right hemisphere: the left is crucially concerned with it, whereas the right is not." If the left hemisphere was "propositional" (a description Bogen borrowed from the nineteenth-century work of John Hughlings Jackson), then the "other side of the brain" was (as Bogen coined it) "appositional."[25] The romancing of the right hemisphere had officially begun (see figure 10).

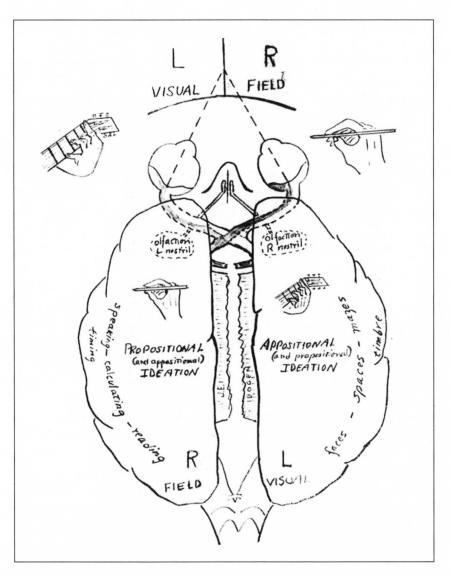

FIGURE 10. In 1975, Bogen illustrated how the right cerebral hemisphere predominates in "appositional ideation" (including visuospatial processing, arts, and music), while the left cerebral hemisphere predominates in "propositional ideation" (including speaking and writing). Courtesy of Meriel Bogen Stern.

Left-Brained or Right-Brained

"Two very different persons inhabit our heads," dramatically opened science reporter Maya Pines in her cover story "We Are Left-Brained or Right-Brained" for the *New York Times Magazine* in 1973. "One of them is verbal, analytic, dominant. The other is artistic but mute, and still almost totally mysterious." Pines declared further that the "sudden surge of interest" in the "nonspeaking side of the human brain" was "probably no accident at a time" when widespread fascination with the nearly "nonverbal disciplines" of yoga and meditation were "enjoying such a vogue." In her discussion of the split-brain research conducted by Sperry, Gazzaniga, and Bogen, Pines wrote, "We know almost nothing about how the right hemisphere thinks, or how it might be educated — and we have just begun to discover how much it contributes to the complex, creative acts of man." Pines noted that persons who wished to improve the rational and literate or intuitive and creative sides of their personalities might soon "learn to activate" their "left or right hemisphere voluntarily." Certainly, Pines observed, recent studies of the split brain were nothing less than "extraordinary in their implications."[26]

In profiling the capacity of individuals to jumpstart the hemispheres of their own brains, Pines cited the research of Robert Ornstein, a psychologist at the Langley Porter Neuropsychiatric Institute at the University of California, San Francisco. Not unlike Bogen, Ornstein had begun to proselytize how the left-hemisphere-dominated West had too long demeaned the right-hemisphere philosophies of the East.[27] Ornstein believed that individuals could practice their right-hemispheric skills and thereby learn effectively to *double* their mental capabilities. As Ornstein wrote in 1972 in his bestselling *Psychology of Consciousness*, people had the capacity to "shift from the individual, analytic consciousness to a holistic mode, brought about by training the intuitive side of ourselves."[28] Throughout, Ornstein preached the psychological *and* neurological benefits of techniques like Zen meditation, Sufism, and yoga. In 1973 Ornstein further solidified his reputation as a sort of whole-brain guru in an op-ed essay for the *New York Times* where he observed that "we can now recognize that we are biologically equipped" with "two separate information-processing systems." He added that since these two modes "complement each other," attempts to get at a "purely 'rational' understanding of the esoteric traditions such as Sufism, Buddhism, Yoga, is not possible since many of the techniques work in the silent language of the right hemisphere — ritual, dance, martial art, specially constructed forms such as temples, cathedrals, contemplation objects, special music."[29] And that same year, Ornstein published an interdisciplinary anthology, *The Nature of Human Consciousness: A Book of Readings*, which promoted how "two major modes of consciousness exist in Man, the intellectual and its complement, the intuitive." To this end, Ornstein freely, or perhaps randomly, juxtaposed articles from the *Archives of General Psychiatry* and *Scientific American* with classic psychological research on perception and split-brain writings by Gazzaniga and Bogen; an excerpt from the *I Ching* and essays on Zen meditation were also included.[30] Bogen may have sought to bring neuropsychology into the counterculture, but Ornstein did more than anyone to raise public awareness about how to unlock whole-brain potential. Ornstein really got the word out and, in the first years of the 1970s, became the right brain's first true evangelist.

Ornstein's theories emerged from laboratory experiments he had conducted on right-hemispheric activation at the Langley Porter Neuropsychiatric Institute. Assisted by psychiatrist David Galin, Ornstein attached electrodes to the right and left sides of the scalps of "normal people" (that is, subjects who had not had surgery to sever their corpus callosum) so that they might record their subjects' EEG patterns while they accomplished various tasks (like writing a letter or constructing a block design from memory). When subjects performed a language task, there was more left-hemispheric involvement; when they performed a spatial task, there was more right-hemispheric engagement. Having demonstrated that EEG results differed depending on the tasks performed, Ornstein and Galin went on to hypothesize that "in most ordinary activities we simply alternate between cognitive modes rather than integrating them," and they concluded with the speculation that ultimately their work might facilitate "the training of ordinary individuals to achieve more precise control over their brains' activities."[31] Again, Ornstein and Galin deduced that a key to turning on the right hemisphere might be meditation. "All meditation techniques are alike in that all involve a focus of attention," observed science journalist Marilyn Ferguson in 1973 in her book *The Brain Revolution*. Citing the research of Ornstein and Galin, Ferguson wrote that "in meditating subjects" there was "the possibility that a new generator, another source of energy, had been activated in the brain." Ferguson added, "Neurologically, the focus [produced by meditation] stimulates the brain rhythmically."[32] Or as psychologist Thomas H. Budzynski recommended in 1976, "a combination of self-directed phrases, breathing exercises, and the new technique of biofeedback" could help individuals "control and manipulate various internal states," which could "promote a quieting of the autonomic functions as well as the skeletal muscle system" and thus in turn "produce a relaxed, quiet, inward-turning state of mind." Building on earlier ideas that the two halves of the brain might be competing for expression, some defenders of the right hemisphere recommended trying to reduce left-hemispheric activity. As Budzynski, for example, optimistically concluded, "Perhaps the decrease in critical, analytical, logical linear-functioning that occurs with a lowering arousal level is the gradual, functional disabling of the major [that is, left] hemisphere."[33] If the achievement of a more purposeful and productive life was a key aim of existence, then the reduction of left-hemispheric activity was likely an excellent means to get there.

Neuropsychological experimentation soon led to announcements that such research could be used to improve the ways in which persons were treated for a range of emotional disorders. Harvard psychologist Gary E. Schwartz reported hopefully in 1975 that "once the basic processes" of "hemispheric asymmetry data" had been "isolated," it would become possible for subjects who suffered from anxiety or depression to be "trained with biofeedback to regulate specific patterns of EEG activity across the hemispheres and to relate these physiological states to specific underlying cognitive and affective experiences." Much like Ornstein and Galin, Schwartz suggested that subjects might well be able to produce "self-induced cognitive states" (such as "a happy state") by learning how to "regulate patterns of physiological activity."[34] The *Saturday Review* too observed in 1975 that techniques geared to "intervening in the brain's processes" could assist individuals "to experience our sensations more joyfully, to keep ourselves mentally and physically healthy, to learn and remember better, to 'know who we are' with a greater sense of wisdom and maturity, to be more creative."[35] Taken together, then, among the various hypotheses derived from neuropsychological laboratory experiments, there was here again a clearly advanced view of the human brain as essentially plastic; individuals possessed the ability to effect long-term alterations inside their own minds.

Cognitive Styles and the Social Order

Advocating that the right hemisphere was a key to inner harmony had undeniable appeal. So too did a notion that the human brain could be reformed and improved. Everything got more complicated, however, when the subject shifted to the potential implications of split-brain research for an under-standing of race, social class, and intelligence. Here a concept of neuroplasticity was less consistently or coherently applied.

The neuropsychological literature of the 1970s and 1980s recurrently stated that there were differences between the races in cerebral specialization. A growing body of empirical research in these decades bolstered claims that minority groups exhibited a right-hemispheric dominance that contrasted sharply with the purported left-hemispheric dominance of whites. In 1977, for instance, the *International Journal of Neuroscience* reported on a neuropsychological experiment that used EEG recordings to map differentials in the right-left hemispheric activity of sixteen bilingual Hopi Indian children. The children first listened to a story in English and then again in Hopi. The study found that the children experienced "a greater right hemisphere participation in the processing of the Hopi speech," leading researchers to propose that the very nature of Hopi language caused children to become more right-hemispheric, while the English language resulted in greater left-hemispheric activity.[36] A study of Navajo college students in 1979 came to a similar conclusion: "Native American Navajo subjects do, in fact, process language in what would normally be considered the minor cerebral hemisphere," and so therefore "it may be that different ethnic groups have been predisposed toward developmental variations in neuropsychological asymmetries."[37] A follow-up study (with Navajo children) published in the journal *Child Development* in 1980 elaborated that the brain of the indigenous person appeared to process language in the right hemisphere due to the fact that the world of the Native American Indian was "more appositional in nature."[38]

Meanwhile, however, neuropsychological literature often explained the presumed right-hemispheric dominance of nonwhite peoples as a result of socioeconomic differences. Already in 1969 Joseph Bogen had concluded that persons labeled "'culturally disadvantaged'" most probably possessed right-hemispheric talents that "remained undeveloped for lack of proper schooling," and he suggested that class background had a profound impact on the brain development of these nonwhite children.[39] Social factors also played a large role in the analyses of UCLA sociologist Warren TenHouten's influential split-brain theories. In 1971 TenHouten received a research contract from the U.S. Office of Economic Opportunity to assist with a study on thought, race, and opportunity.

In his contribution, which he titled *Cognitive Styles and the Social Order*, TenHouten applied "neurological theory of lateral specialization for verbal and visual brain functions" to a neuropsychological study of African Americans, Hopi Indians, and whites. TenHouten found, for instance, that while the Hopi considered language to be "a means of observation, a vehicle for the discovery of ever-emerging levels of vibrations, and pre-eminently natural phenomenon," English-speaking white persons took language to be "a means of expression, a vehicle for argumentation, and a pre-eminently social phenomenon." TenHouten went on to argue that these linguistic differences reflected different "cognitive styles," which originated in differences in hemispheric dominance; these differences in hemispheric dominance thus translated into "differences in group access to modes of thought and socioeconomic rewards" within a Western left-brain-dominated social order. In this way much like Bogen (with whom TenHouten had collaborated), TenHouten did not take "cognitive styles" to be a result of *innate* differences between nonwhite and white persons in brain lateralization. Rather, he proposed that minorities — and perhaps especially African Americans — developed "cognitive styles" distinct from whites due to the black community's "denial of access to the resources of white society." TenHouten was insistent that his aim was to "undermine . . . racist thinking." While whites used speech in "propositional" ways, "black speech" was by contrast more "appositional" and "places more emphasis on metaphor." Further, although "black children do not learn to present their thoughts as a linear sequence of statements," black language was "probably far richer in its nonverbal dimensions." What his meta-analysis of the extant research showed was that one could see among whites and nonwhites "two independent modes of cognitive performance, and the definition of one as 'intelligence' represents a value judgment, and nothing more than that." It was indefensible that "the American educational system and the economy reward propositional performances more than they do appositional performances." For, TenHouten stressed, "both are 'intelligences' in and of themselves."[40]

Neuropsychological experiments soon provided what was believed to be additional empirical support for a hypothesis that race, socioeconomic status, and cerebral specialization could be correlated. For instance, when neuropsychological researchers in 1976 used a dichotic listening test (premised on the idea that since the left ear corresponded with the right hemisphere, a full appreciation of emotional tones and statements was likely to be processed principally through the left ear rather than the right) on white and black children of various socioeconomic backgrounds, they determined there were no innate "racial differences" in the development of cerebral dominance. Instead they found "significant social class differences" in how these children's brain hemispheres developed. These researchers concluded that their findings underscored that an uneven development of a child's brain was probably caused by "socio-cultural variables."[41]

As Bogen had already argued in 1969, others now came also to underscore how the heightened vulnerability of the left hemisphere to environmental stresses in early childhood meant that low socioeconomic status could likely result in a right-brain dominance. Whatever *differences* there were imagined to be between white and nonwhite persons were definitively not meant to be interpreted as *deficits*. On the contrary, a consistent refrain in all neuropsychological research on "cognitive styles and the social order" was that the right-hemispheric dominance said to be exhibited by members of minority groups (such as Hopi Indians and African Americans) was likely a source of special skills and abilities often described as sorely lacking among white people. Nonetheless, inevitably, even as environment and economics were invoked, assumptions about correlations between racial-ethnic groups and cerebral specialization remained.

Such assumptions regarding the right-hemispheric dominance of non-white students cannot be understood apart from the context of the ongoing wrenching controversies concerning lagging educational achievements of these same nonwhite students. As mentioned at the outset of this chapter, Berkeley psychologist Arthur Jensen was asserting that African American children *as a group* did not and *would never* be able to match the intelligence quotient of white children and that the cause of this differential was largely the result of genetics. In 1969 Jensen declared that it could come as no great shock that "the chief goal of compensatory education — to remedy the educational lag of disadvantaged children and thereby narrow the achievement gap between 'minority' and 'majority' pupils — has been utterly unrealized in any of the large compensatory education programs that have been evaluated so far." Jensen further argued that "the uniform failure of compensatory programs" made sense once it was acknowledged that heredity (and not environmental factors) *principally* determined individual intelligence — and that the genetic makeup of African Americans likely dragged down their average IQs. Jensen preemptively observed that while his views would inevitably face resistance, such resistance did not mean he was incorrect in his conclusions. Instead, Jensen provocatively observed, "The idea that the lower average intelligence and scholastic performance of Negroes could involve, not only environmental, but also genetic factors has indeed been strongly denounced.

But it has been neither contradicted nor discredited by evidence."[42] As Jensen anticipated, his remarks did provoke outrage from psychologists and educators committed to programs that provided assistance to impoverished children of color.[43]

Commentators like TenHouten and Bogen who understood themselves as ardent anti-racists certainly had no love for Jensen's theories concerning race, social class, and IQ. But they tended to take Jensen's arguments as further (if accidental) evidence to support their own neuropsychological research on the right-hemispheric dominance of nonwhite and non-Western peoples. In 1971 TenHouten specifically distanced himself from Jensen's racialist notion that genetics were at the root of lower IQ results for African American children, even while he also spun Jensen's view to foreground that what IQ tests demonstrated was certainly *not* that genetics chiefly determined intelligence. Rather, the key for IQ tests was that they should become more attentive to differences in brain function. TenHouten pointedly declared, "It is hypothesized that the direction of causality works the other way, i.e., social forces may be at work which cause some groups to specialize in left-hemisphere thought and other groups in right-hemisphere thought."[44] And in a like fashion, Bogen (in the final installment of his treatise "The Other Side of the Brain," published in 1972) also cited Jensen's writings on IQ, genetics, and race not in order to provide evidence for a genuine gap in intelligence between races but rather in order to promote the view "that subdominant groups in a technological society are provided less access" to left-hemispheric skill-building and "consequently must rely more often upon the alternative" of right-brain talents. Bogen observed as well that "certain ethnic groups" were the beneficiaries of "relatively greater development of right hemisphere potential" and proposed that the "hemispheric specialization" of indigenous peoples and persons of color was due to the fact that these peoples had grown up in subcultures that emphasized "spatial skills" and deemphasized "intensive education of the left hemisphere potential for reading, writing, grammar, etc." Thus the "cultural differences" between white people and persons of color "can be interpreted in part as a result of asymmetry in hemispheric utilization." Strikingly, Bogen elaborated that so far as he was concerned, "the nature-nurture issue is peripheral to our main point" because it hardly mattered whether "a right-hemispheric skill was innate or learned," precisely because "either explanation" was "compatible with our suggestion that cultural differences can be interpreted in part as a result of asymmetry in hemispheric utilization."[45] That an IQ test score was an inaccurate measure of intelligence due to a left-hemispheric bias that neglected to take into account a child's whole range of cognitive attributes was taken as a given. The differential in test scores between white and nonwhite children proved this. Bogen's logic was as clear as it was circular.

Nor did Bogen ever manage to settle the puzzle of how best to rebut derogatory views on the assumed connections between race and intelligence, even while he remained preoccupied with it. In 1975 in "Some Educational Aspects of Hemispheric Specialization," Bogen railed against how "the usual justification of IQ tests is that they predict further scholastic achievement, and that the latter is in turn predictive of 'life success.'" He added that efforts to predict "'life success'" were "ultimately based upon a criterion of 'success' which is not only most often measured monetarily, but seems to depend in part upon an analytic attitude hypertrophied by centuries of contention against nature." The reason so many students chose to reject traditional approaches toward education, Bogen noted, was that conventional pedagogical methods solely addressed students' "left-hemisphere potential," while they typically disregarded students' desire to "learn to live within nature as bilaterally educated, whole persons."[46] Yet here again Bogen ran into a paradox, as he said that IQ tests provided an *inaccurate* measure of true intelligence while simultaneously contending that these tests demonstrated *real* neurological differences between racial groups. And in 1976 psychologist Andrea Lee Thompson and Bogen again turned specifically to data from IQ tests "administered to urban samples of black females, black males, white females, and white males as well as to Hopi Indians and rural whites" to argue that any differences in scores were the result of "cultural differences"— and thus the result of right-hemispheric dominance. A lower IQ among nonwhite groups (as well as many poor and rural whites) could be the result of damage done to the left hemisphere due to early socioeconomic deprivations, they contended.[47] In sum, all through, Bogen worked passionately to use split-brain research to challenge a repressive left-hemispheric (and Western and white) social order.

Among neuropsychologists and the many advocates of their research on cerebral asymmetry, it was a simple fact that African American children had a right-hemispheric dominance, whether due to environmental or biological factors (it apparently did not matter), and thus possessed superior skills in holistic perception, nonverbal reasoning, and music — skills IQ tests ignored. Jensen's evidence may have been accurate: black children did lag behind white children on IQ tests. But Jensen's interpretations were wrong because they had not taken into account neuropsychological research on cerebral dominance. Jensen argued that there existed a correlation between intelligence, race, and genetics. He should have argued that extant intelligence testing be revamped to incorporate the cerebral specializations of *all* children of *all* racial and socioeconomic backgrounds.[48]

By the later 1970s and early 1980s, it became a commonsense consensus that the right hemisphere of countless minority and disadvantaged young person's brains was getting short shrift in the setting of the American classroom and that gross disinterest in right-hemispheric learning was causing these same children of color to flounder and fail. Indeed, it was "the dominant white culture" that had "almost religiously accepted the rational logical modes of the left hemisphere," as psychologist Bob Samples wrote in 1976 in his popular right-brain manifesto, *The Metaphoric Mind.* Thus, Samples added, "minorities, locked out of the contexts of the dominant culture, have been falsely judged to be inferior in motivation."[49] Or as an expert in child development inquired in 1980, "What is our loss when schools stress the more measurable left hemispheric mathematical and verbal skills which are referred to popularly as 'the basics' and at the same time ignore the development of 'right brained' intuitive thinking?"[50] And African American psychologist Gerald Gregory Jackson forcefully underscored a comparable point in 1982, noting that since "the psyche of the Western world" was "dominated by the left hemisphere of the brain" and that black people living in the West possessed a "right hemisphere" capacity to combine "European rationality with African intuitiveness," discriminatory practices in schooling were almost inevitable.[51] The new goal was declared to be hemispheric equality — not further neglect.

Right-Brained Kids in Left-Brained Schools

Decrying the injustices of an educational system geared to left-hemispheric (that is, white and middle-class) students became ever more fashionable in the course of the 1970s. Already in 1973 Maya Pines had dramatically concluded that a growing body of neuropsychological evidence indicated how "children from poor black neighborhoods generally learn to use their right hemisphere far more than their left — and later do badly on verbal tasks."[52] Ornstein and Galin delivered a paper at the annual meeting of the American Association for the Advancement of Science in 1974 in which they applied their research on brain lateralization to explore the relationship between the limited educational opportunities available to right-hemispheric students and socioeconomic inequities. Much like TenHouten and Bogen, Ornstein and Galin argued "that subcultures within the United States are characterized by differences in predominant cognitive modes: the middle class more often employ the verbal-analytic mode, the urban poor are more likely to use the spatial-synthetic mode." They added, "This could result in a cultural conflict of cognitive style and may in part explain the difficulties of the urban poor children in a school system oriented toward the middle class. There seems to be a new recognition among educators of the importance of both modes of experiencing the world."[53] In 1975, psychologist Michael Gazzaniga weighed in, cautioning, "When a child's talents lie in visual-spatial relations, and he or she is being forced into a curriculum that emphasizes the verbal articulatory modes of solving a conceptual problem, this child will encounter enormous frustration and difficulty, which may well result in hostility toward the teacher and, worse, toward the learning process itself."[54] Praising the right hemispheres of black and poor children (whether implicitly or explicitly) as a means to counter the perception that these children were in any way lesser became a recurrent thematic in the neuropsychological literature of the 1970s.

No less an authority than Roger Sperry came also to advocate for the reform of educational policies on the basis of ongoing research on hemispheric dominance and cerebral specialization. Repeatedly in the 1970s, Sperry argued that the right-hemispheric skills of children were being neglected by the left-hemispheric emphases of American curricula. In 1973 Sperry put the matter succinctly: "What it comes down to is that modern society discriminates against the right hemisphere."[55] In 1975 Sperry emphatically underscored this point when he opined how "our educational system and modern society generally (with its very heavy emphasis on communication and on early training in the three Rs) discriminates against one whole half of the brain."[56] Sperry even used the occasion of his lecture in Stockholm on being awarded the Nobel Prize in 1981 for his research on the bisected brain to comment on how "the need for educational tests and policy measures to identify, accommodate, and serve the differentially specialized forms of individual intellectual potential" had become "increasingly evident."[57]

No doubt encouraged by pronouncements from neuropsychological experts, specialists on education looked increasingly to split-brain research for guidance on how to reconceive the learning experiences of students. When education psychologists at Stanford University in 1975 examined a student's decision on where to sit in a classroom (on the left or the right side of the room), they determined that this decision might be a predictor of that student's left- or right-hemispheric dominance (with analytic left-brain students displaying a preference for the right side and right-brain students selecting the left side of the room).[58] When the National Education Association of the United States in 1976 announced how urgently important it was that schools address the special needs of "right-brained kids" trapped "in left-brained schools," it concluded that it was unacceptable that children be "handicapped" in their studies simply because they were actually "more proficient in right hemisphere (visual) input processing."[59] And when psychologist (and leading authority on how to nurture creativity in children) E. Paul Torrance argued that there existed "a rather consistent tendency for the measures of creative style to be positively and significantly related to the right hemisphere style of information processing and negatively and significantly related to the left hemisphere style," he indicated that it would be a gross error to pressure the most imaginative and right-hemispheric children to endure more left-hemispheric pedagogical approaches.[60]

Such concerns came as well to have an impact on the field of special education. The *New York Times* cautioned in 1978, for instance, that "the brain's division of labor" meant that right-brain schoolchildren were being unfairly tracked into remedial classes. "Given new insights into the nature of the two hemispheres of the brain," the *Times* critically observed, "educators should be wary about assuming that they are dealing with a slow learner."[61] And Harvard neurologist Norman Geschwind questioned whether the classification of "learning disabilities" really made any sense since such labels simply revealed the biases of a highly rational and literate society that lacked an appreciation of both hemispheres of the brain. Geschwind wrote in 1978 of "the child who has trouble learning to read": "My suspicion would be that in an illiterate society such a child would be in little difficulty and might, in fact, do better because of his superior visual-perceptual talents, while many of us who function well here might do poorly in a society in which a quite different array of talents was needed to be successful."[62] Likewise there were several studies that concluded how children's "right hemispheric reading strategies" might cause them to read more slowly than children who read with "left hemispheric strategies."[63] Educators speculated that the student with a reading impairment could not "lateralize" properly and that this child's right hemisphere was "being used for a job it is poorly equipped to handle." If such was the case, the *Journal of Learning Disabilities* concluded in 1977, "all the federal funding and suitable texts in the world won't help the child who is inadvertently using the wrong side of his brain to do the job" of reading a book.[64] The journal *Topics in Learning and Learning Disabilities* put the question bluntly in 1983: "Are dyslexics locked into the right hemisphere?" (The answer was an emphatic *yes*.)[65]

Cerebral Asymmetry and Intelligence Testing

By the late 1970s, education and school psychologists began articulately to argue that what intelligence tests measured was not in fact intelligence but rather an individual child's ability (or lack thereof) to integrate his or her two distinct modes of processing information — analytic (left hemisphere) with holistic (right hemisphere). Cerebral asymmetry meant that one processing mode was dominant (whether left- or right-brain), and educational psychologists suggested that an intelligence test principally assessed only the dominance of a child's right or left hemisphere — and not a child's whole-brain intelligence. Tests that did not take a child's cognitive style of learning into account could not offer a thorough assessment of that child's cognitive capabilities.

No one did more to redirect debate over intelligence testing in the 1970s and 1980s with the aid of neuropsychological research on the split brain than school psychologist Alan S. Kaufman. Kaufman was well positioned to make his mark on the field of intelligence testing. He had studied with Robert L. Thorndike, a leading psychometrician at Columbia University, and had worked closely for several years with David Wechsler, the psychologist responsible for the Wechsler Intelligence Scale for Children (WISC), first developed in 1949 and then revised in 1974 (as WISC-R). Kaufman thus had an extensive background in measurement research and intelligence testing, and his view by the late 1970s was not only that IQ tests tended overwhelmingly to lack an up-to-date incorporation of neuropsychological theory but also that intelligence testing often proved to be insensitive to the learning processes specifically of minority children. For instance, Kaufman wrote in 1979, as he reviewed close to a decade of neuropsychological research on hemispheric specialization, that "current individual intelligence tests" were not working because they "do not begin to tap the rich resources of the right cerebral hemisphere" and so discriminated against "socioeconomically disadvantaged black children" with "a right brain leaning." Kaufman went on to echo the position of Joseph Bogen (whose writings he not infrequently cited) that "the adaptability of the left hemisphere may also make it far more vulnerable than the right hemisphere to the effects of cultural deprivation. Socioeconomically disadvantaged black children may therefore have a right brain leaning not only because of the stress on gestural communication within their culture, but because of its possible greater resilience in the face of deprived circumstances." And Kaufman asserted, "The possibility that blacks may have some degree of preference for the right hemisphere may actually suggest an inhibiting effect on their scores earned on left-brain tests."[66] From here it was only a small step to making an emphatic anti-racist case against all existing intelligence tests on the grounds that real differences existed (due either to environmental damage or prewired difference) in the ways the lateralized brains of white and African American children processed information.

In his now-classic 1979 text, *Intelligent Testing with the WISC-R*, Kaufman made neuropsychological research on hemispheric specialization central to his argument calling for the reform of IQ measurements. "A mounting body of evidence has shown the left hemisphere of the brain to specialize in sequential processing, which is analytical and successive in nature, and to be particularly well suited for handling linguistic and numerical stimuli," Kaufman wrote, adding that "in contrast, the right hemisphere (long the forgotten part of the brain) excels in multiple processing, which is holistic and simultaneous in nature, and handles visual-spatial and musical stimuli with facility." And here Kaufman quickly brought in the issue of race: "The underrepresentation of right-brain assessment in intelligence tests may be particularly penalizing to black children." Kaufman continued: "Some cultures may be left brained and verbal in their orientation, whereas others are right brained and nonverbal. The emphasis on music and movement among blacks implies a right-brain preference, as does the central role attributed to nonverbal and gestural communication within the black community." Elsewhere in the same text, Kaufman wrote, "The existence of a possible right-brain leaning within the black community may render the limited coverage of right-hemisphere processing by the WISC-R especially penalizing to black children. Indeed, a relative dependency on the right hemisphere may characterize not only blacks but most 'urban poor' groups as well." Adopting wholesale the notions of Joseph Bogen and David Galin, among others, Kaufman restated what he called a "right-brain resilience hypothesis":

The right hemisphere seems to be more mature than the left at birth, both physiologically and functionally, and is also a more pervasive force in the early stages of life. An infant perceives and learns nonverbally, sensorily, and spatially to a large extent during the first year of life, styles of learning that are congruent with the processing modes of the right cerebral hemisphere. Although less mature than the right brain, the left brain is more adaptable at birth and has the capacity of sub-suming complex and analytic functions. The greater adaptation of the left hemisphere may conceivably render it unusually vulnerable to the impact of cultural deprivation. Hence disadvantaged children may have a right-brain leaning, at least in part, because of the resilience of the right hemisphere in the face of deprived environmental conditions.[67]

By 1980, Kaufman stated his argument about race and intelligence testing even more directly: "The cognitive processing models steeped in neuropsychology have many implications for the evaluation of the intellectual abilities of minority-group children," especially because "blacks seem to utilize right-hemisphere intelligence in their everyday lives, but these skills are neither tapped nor rewarded by our contemporary intelligence tests." Therefore, as Kaufman dramatically concluded, the IQ yielded by "intelligence tests should be treated as incomplete measures of children's global intelligence, particularly for blacks who may (as a group) tend to have a right brain preference."[68] In this way, the call to reformulate intelligence testing — one embedded both in neuropsychological theory *and* in a self-understood-as-anti-racist sensibility — came largely to engage a notion (whether innate or culturally induced, it was never precisely clear) of racial difference.

It was a call Kaufman himself answered, developing in these same years (together with his wife, Nadeen Kaufman) the Kaufman Assessment Battery for Children (the K-ABC), a new clinical test to measure the intelligence of children as young as two and a half years — and a test that specifically sought to integrate split-brain theory into its methodology. A central aim of the K-ABC was successfully to minimize test score differences between minority and white children; the Kaufmans sought to accomplish this goal through an attentiveness specifically to what they assumed were minority children's dominant right hemispheres.[69] To choose a crucial example, the K-ABC used scales for simultaneous (right-hemispheric) processing in addition to the traditional scales for sequential (left-hemispheric) mental processing. And the test's results were impressive; the intelligence gap on K-ABC tests between white and nonwhite children did indeed narrow significantly—a result Kaufman saw as a sign of the test's "cultural fairness."[70]

Arthur Jensen, a man who had staked his professional reputation on the argument that there existed significant average differences in intelligence between white and black children, was underwhelmed by the K-ABC test results. Jensen immediately condemned the K-ABC as a major step in the wrong direction. He declared the "diminished black-white difference on the K-ABC" to be nothing but "the result of psychometric and statistical artifacts." And he concluded therefore that the K-ABC was a "more diluted and less valid" measure of intelligence than prior tests that demonstrated black children to be intellectually inferior to white children.[71] Jensen's argument against the K-ABC would be recycled more or less verbatim a decade later in psychologist Richard J. Herrnstein and political scientist Charles Murray's *The Bell Curve*, which underscored that when the Kaufmans' test demonstrated a "diminished gap" between black and white test results, this "largely reflected psychometric and statistical artifacts."[72] Undeterred by such blistering attacks, Alan Kaufman and clinical psychologist Elizabeth O. Lichtenberger confidently responded that Jensen's evidence "does not support his own point."[73]

In revised form, the K-ABC has endured. In 2005 Kaufman and Lichtenberger announced the KABC-II, a "structurally and conceptually" transformed assessment test, even as they reiterated how the first K-ABC's "fairness in assessing children from diverse minority groups made it stand out above other tests." But by this point in the early twenty-first century, and despite the KABC-II's stated reliance on a "neuropsychological model," any reference to Joseph Bogen or right-brain resilience or anything at all related to how racial difference might be traceable to differences in cerebral dominance had been all but completely expunged from the assessment battery's historical narrative.[74]

The K-Mart of Brain Science

Not everyone concurred that split-brain research had—or should have— broad cultural, educational, or political implications. The *New Yorker* already in 1977 grabbed a chance to parody the impact of split-brain theory on the corporate world (see figure 11). A handful of psychologists argued that the claims being made by split-brain advocates were distortions of real science. A small backlash against the split-brain paradigm took shape. Cognitive psychologist Robert D. Nebes, who had trained with Sperry at Caltech, complained already in 1975, "Lately, everything from creativity and imagination to the id, ESP, and cosmic consciousness have been suggested to reside in the right hemisphere." It was, Nebes added, as if the right hemisphere had been annexed "by counterculture groups as their side of the brain."[75] In 1977 psychologist Daniel Goleman (despite the fact that he actively promoted split-brain theory elsewhere) criticized "the hemisphere fad" because it fomented "widespread confusion between the poetics of experience and the hard facts of brain function."[76] Developmental psychologist Howard Gardner expressed weary despair at split-brain mania. "It is becoming a familiar sight," Gardner wrote in 1978, sardonically adding, "Staring directly at the magazine reader — frequently from the publication's cover — is an artist's rendition of the two halves of the brain. Written athwart the left cerebral hemisphere (probably colored in stark blacks or greys) are the words 'logical,' 'analytical,' or 'Western rationality.' More luridly etched across the right cerebral hemisphere (perhaps in rich orange or royal purple) are the words 'intuitive,' 'artistic,' or 'Eastern consciousness.'"[77] The *Journal of School Psychology* went out of its way derisively to declare in 1979 that there was "no scientific basis" in neuropsychological research "for any reorganization of curricular, teaching, or testing programs within contemporary educational practice."[78] And an account from the mid-1980s drily noted — in near-incompatible metaphors — that the "widespread cult of the right brain" had pulled down "the duplex house that Sperry built" so that there could be constructed in its place "the K-Mart of brain science."[79]

Others who had been active in promoting neuropsychological research on cerebral specialization began to speak out more generally against dichotomania. Psychologist and neuroscientist Michael C. Corballis proposed already in 1980, "This transcendental conception of hemispheric duality is inspired more by age-old myths about left and right than by the empirical evidence."[80] Neurologist Joseph E. LeDoux, who had studied with Gazzaniga, wrote in 1983 that while "cerebral asymmetry has attracted much attention in recent years," it was nevertheless an unfortunate fact that "speculation concerning the implications of hemisphere differences has gone well beyond the data."[81] By the middle of the decade, Gazzaniga had himself become more harshly dismissive. "It's pure nonsense," he reflected about claims concerning the talents of the right hemisphere. He sarcastically added, "The simple fact is that you don't have to invoke one cent's worth of experimental psychological data or neuroscience to make the observation that there are

"Foster here is the left side of my brain, and Mr. Hoagland
is the right side of my brain."

FIGURE 11. A *New Yorker* cartoon by Lee Lorenz in 1977 parodies the degree to which split-brain theorizing had come also to inform beliefs prevalent in the corporate world. Reprinted by permission of the Cartoon Bank/Condé Nast.

some people in this world who are terribly intuitive and creative, and some who aren't."[82] Additionally, over time, neuropsychological experiments failed to replicate many of the more exuberant results linked to the bisected brain. One study challenged the reliability of dichotic listening experiments concerning asymmetry in hemispheric responses to musical tones and verbal speech; the study argued that "ear preference" was often arbitrary, and it calculated that 30 percent of subjects altered ear preference when retested.[83] A study in the *American Journal of Psychiatry* "did not provide support for the notion" of "structural asymmetries in schizophrenic patients," thus casting doubt on Julian Jaynes's popular theory of madness and "the bicameral mind."[84]

Split-brain research that suggested that minorities and white people or women and men exhibited differential brain activation also did not prove to be especially robust. A study in 1981 concluded that "both the direction and magnitude of language laterality appear to be identical in Anglo and Na-vajo persons."[85] A study conducted with bilingual Crow Indian children did not establish that these children's right hemispheres were more dominant.[86] After Native American scholar Roland Chrisjohn and neuropsychologist Michael Peters examined all the available split-brain studies on indigenous hemispheric brain activity, they angrily wrote that the idea of "the right-brained Indian" represented nothing more than a "pernicious myth."[87] The claims advanced by psychologist Jerre Levy and others that the female brain might suffer from a deficit in spatial abilities (whereas the male brain ex-celled at these) were met with strong skepticism as well. In a review of Levy's research, biologist Anne Fausto-Sterling concluded in 1985 that while con-cepts about the split brain and gender difference remained "still in fashion," there was in truth "no evidence" to support them — as in "quite simply, none whatsoever."[88]

Teaching and the Human Brain

Yet despite these diverse disclaimers and outright refutations, split-brain theory was just coming into its own in the discipline of education. Edu-cators cited split-brain research (including dichotic listening tests) in their defense of the neuropsychological benefits of bilingualism, just as many public school systems were shifting away from their support for bilingual programs. The argument was that learning two languages strengthened the whole brain.[89] Already in the late 1970s, but through the 1980s and accelerat-ing after the publication in 1991 of psychologist Renate Caine and Geoffrey Caine's highly influential *Making Connections: Teaching and the Human Brain*, leading educators and psychologists articulated a comprehensive case for applying split-brain research to curricular reforms in actual educational settings.[90] The fine arts (for example, art, music, theater) were among the ini-tial scholarly areas where split-brain research had an often profound impact on pedagogical practices, spurred in large part by the publication of Betty Edwards's *Drawing on the Right Side of the Brain*.

At a moment when budgets for the fine arts were being slashed, neuro-psychological evidence about the distinctive way the right brain processed the world became a useful weapon. A view that right-brain-stimulating activities might be used to assist in the teaching of (the purportedly left-hemispheric-grounded matters of) language and literacy followed, as educa-tors contended that especially slow learners could be helped through the use of body movement and the incorporation of visuospatial techniques into the instruction of reading and writing skills.[91] Elementary and high schools from New Jersey to Michigan to California soon began to implement what they called "brain-based learning"— and school principals cited strong successes after doing so.[92] By the early 1990s, there was much talk as well in education circles about psychologist Frances Rauscher's so-called Mozart effect, that is, the cognitive benefits that appeared to result after very young children listened to a Mozart sonata.[93] Rauscher's argument was that when children listened to classical music, it helped to "systematize the cortical firing patterns so they can be maintained for other pattern development duties, in particular, the right hemisphere function of spatial task performance."[94] (By the late 1990s, Governor Zell Miller of Georgia was citing the Mozart effect when he proposed a state budget that allocated funds to provide a cassette tape or compact disc of classical music to mothers of newborns when they came home from the hospital.)[95] Also, the potential advantages of "whole-brain learning" for the academic performance of African Americans once again became a theme in educational discourse.[96] In every instance, the proponents of brain-based curricular reform announced a desire that students be taught "in accordance with the way the brain is naturally designed to learn."[97]

An agenda that tapped the potential neuroplasticity in underprivileged children has continued to undergird much twenty-first-century discussion of brain-based education reform. If the brain can be rewired in the context of an enriched educational environment, or so the argument goes, then *neuroeducation* — with its call for a more holistic pedagogical approach — represents a powerful rebuke to a single-minded obsession with math and reading testing scores.[98] In this way, neuroeducation could well be said to be a stepchild of right-brain advocacy a generation earlier. Just as the pronouncements of right-brain partisans from the 1970s were, so too the more recent assertions about the brain and learning put forth by self-styled "neuroeducators" have been prominently debunked as unsubstantiated and overstated.[99] And yet notwithstanding such critiques, many popular pedagogical models for holistic learning — such as those offered by the global network of Waldorf schools — continue to be represented as "provid[ing] a developmental framework aligned with the . . . neuroscience understanding of the development of specific systems of the brain."[100] To keep pace with neuropsychological progress, many key assumptions about hemispheric specialization have since been adapted to be more attentive to the *localization* (not the *lateralization*) of skills (for example, memory and higher cognitive abilities) in specific regions of the brain. Nonetheless, it is clear that the history of split-brain research remains a crucial precursor in terms of its ability to speak powerfully to a broader public. In this redefined context, theories about how to use neuropsychological research — however problematic it may have always been — to optimize the way children learn still find a receptive audience among educators committed to restructuring both school curricula and intelligence testing.

Conclusion

Widespread fascination during the 1960s and 1970s with split-brain research may be taken as yet one further instance of what David Kaiser and Patrick McCray have labeled "groovy science."[101] The allure of split-brain theories may also be interpreted as part of the broader "anti-modernist lament" during and immediately after the sixties, as Anne Harrington has identified.[102] However, the dynamics in this instance represent an even more complicated confluence of countercultural impulses and neuropsychological insight. Neuropsychology offered new evidence for perspectives that were already widely held. The heyday of split-brain research arrived precisely at the moment of "the notorious 'Zen boom'" (as the popularizer Alan Watts labeled it) — a moment primed to identify neurological findings through a countercultural lens.[103] Watts had himself announced a need for "a cross-fertilization of Western science with an Eastern intuition," and research into the bisected brain appeared very much to be an answer to that need.[104] Similarly, as another sympathetic account summarized, a belief that "the Eastern person interprets the world as whole/one, while the Western person interprets it as unjointed details," had been in circulation among Zen proponents long "before split-brain research became popular" in the United States.[105] According to this frame of reference, what the fruits of split-brain research confirmed was the esoteric wisdom of ancient non-Western philosophies.

At the same time, neuropsychological research on the bisected brain came quickly to reshape and redirect debate over how best to reformulate educational practices and how equitably to reconceptualize the intelligence testing of children. In this way, split-brain research emerged — however counterintuitively — as a powerful cudgel with which to challenge racialist arguments that African American as well as other minority children were in any sense inherently inferior. Profiling these children's "intuitive," "spatial," and "nonverbal" talents became one strategic and, to many, persuasive — if also almost always elusive — means of pushing back against an assumed "Western" and "white" definition of intelligence that critics asserted repeatedly in the 1970s and 1980s had come to dominate U.S. society. In this way and provocatively (if unexpectedly), a history of split-brain research can be read as integral to an analysis of the racial history of these decades. Even as the split-brain revolution receded in the course of the later 1980s — only to be supplanted in the 1990s by an even more fungible concept of "emotional intelligence" (itself, as it happens, a racialized concept) — the cultural and practical impact of split-brain theory on educational policies and intelligence testing practices has persisted.

Nonetheless, it is additionally evident that fervent interest in neuropsychological research in the 1970s and 1980s was likely also a characteristic reaction of excitement, overpromising, and increased funding flows in response to a new set of discoveries. It is possible that the best explanation for the fascination with split-brain research was that it provided scientific justification for sociocultural agendas. The split-brain fad — as already noted — represented an important antecedent to the more recent trend that correlates brain regions to cognitive functions, emotions, and behaviors. But split-brain thinking succeeded not least of all because it had something to offer everyone. It had an undeniable attraction for those who believed that brains were plastic and that training could result in transformation and improvement. It attracted the financial bottom-line-oriented business elite. It inspired idealistic do-gooders and anti-rationalist romantics of all stripes. It appealed to those who accepted that there were differences (whether genetic or environmental) between different ethnicities or racial groupings but wanted to give positive valuations to those typically underprivileged groups of people. As flawed and ultimately problematic as neuropsychological research of the split brain was subsequently declared to be, and however misleading (not to mention empirically absurd) the reinforcement of ideas of cerebral asymmetry and racial difference also was, the political and social — and anti-racist — claims made in its name (not least concerning the benefits of a more well-rounded education) proved to have lasting consequences. Important outcomes are sometimes based on imperfect science.

FOUR

A Racial History of
Emotional Intelligence

N 1995, WHEN psychologist Daniel Goleman published *Emotional Intelligence*, a treatise that topped best-seller lists for more than a year and went on to sell five million copies worldwide, he introduced a popular audience to a concept that had been circulating in psychology circles for some time. The idea that people possessed an "emotional intelligence" (EI) was not original with Goleman. There had already been research by several psychologists, including Howard Gardner, Peter Salovey, and John D. Mayer, who had mapped out a position that intelligences were multiple and that noncognitive skills like self-awareness, motivation, and empathy played as large a role — if not a far greater one — in how a person's life turned out as did that individual's IQ.[1] Not incidentally, psychologists and educators who championed the centrality of noncognitive skills were posing a direct challenge to experts across the political spectrum who believed in the value of IQ as a metric — and to those theorists convinced that traditional IQ testing represented the most accurate predictor of a person's capacity for achievement in life. Thus the ascent of a concept of emotional intelligence in the mid-1990s proved most timely as it imparted a novel and powerful (implicitly anti-racist) alternative to a view that cognitive intelligence trumped all other aptitudes — especially in the wake of (and fierce controversy surrounding) the publication of Richard Herrnstein and Charles Murray's *The Bell Curve: Intelligence and Class Structure in American Life* in 1994.

A history of emotional intelligence cannot be told apart from a history of race and class in the United States. This has everything to do with one further dimension of EI, one that has gone underacknowledged in discussions of EI and of Goleman's definition of it. It involves the fact that Goleman's book placed so strong an emphasis upon self-control as a key component of EI. In all the laudatory reception of the book, no one thought to interrogate the genealogy of that emphasis. Yet by tracing the complex — often obliquely coded but nonetheless palpable — interconnections between debates about the value of self-control among psychologists, criminologists, education experts, and policy makers since the 1930s, the inextricability both of EI and of the lessons drawn from talking about it from the racial and class politics of the United States comes into view rather more clearly. So too do the shifting beliefs about individual personality and perfectibility — and about possibilities for social change — in which these debates were embedded.

Notably, the emphasis on self-control had not been shared by earlier proponents of the concept. Rather, as Salovey and Mayer had written in 1990, emotional intelligence was a much broader and diffuse "set of skills" that included "the accurate appraisal and expression of emotion in oneself and in others, the effective regulation of emotion in self and others, and the use of feelings to motivate, plan, and achieve in one's life."[2] But Goleman recast EI as having almost everything to do with an ability to exercise self-discipline. It was a point he hammered in *Emotional Intelligence*, writing, for instance, that "a sense of self-mastery, of being able to withstand the emotional storms that the buffeting of Fortune brings rather than being 'passion's slave,' has been praised as a virtue since the time of Plato." He added, "There is perhaps no psychological skill more fundamental than resisting impulse." He wrote elsewhere in *Emotional Intelligence*, "The bedrock of character is self-discipline; the virtuous life, as philosophers since Aristotle have observed, is based on self-control." He put the good that came with achieving self-control in terms that were quite dramatic: "The capacity to impose a delay on impulse is at the root of a plethora of efforts, from staying on a diet to pursuing a medical degree."[3] Goleman left little doubt as to which noncognitive skill he most associated with emotional intelligence.

To make the case that self-control represented a crucial — if not *the* crucial — aspect of emotional intelligence, Goleman relied almost entirely on a (then) not-well-known series of experiments with children that had been conducted by a Stanford personality psychologist named Walter Mischel. At various points dating back to the 1950s, Mischel had given children a basic choice. Did the child want the smaller notebook now or a larger notebook a week later? Would she accept the smaller magnifying glass immediately or wait for a bigger magnifying glass a week later?[4] Or — as the test span shortened over time — would the child choose to gobble up one marshmallow right away, or would he be able to wait fifteen minutes for the tester to re-turn so that he could enjoy two?[5] Regardless of the rewards involved and the details adapted — and they were adapted frequently over the decades — the core of Mischel's experimental design remained much the same.[6] For in-stance, a report on the first of these gratification delay experiments (con-ducted in 1957) described how a researcher presented young children with a larger and a smaller piece of candy and then announced, "I would like to give each of you a piece of candy but I don't have enough of these (indicating the larger, more preferred reinforcement) with me today. So you can either get this one (indicating the smaller, less preferred reinforcement) right now, today, or, if you want to, you can wait for this one (indicating) which I will bring back next Wednesday (one week delay interval)."[7] Each child then had to decide what to do next.

Importantly, as the published reports growing out of Mischel's experi-ments piled up, Mischel began to identify an ever-widening array of cor-relations, all connected to the single decision of the child to yield or not to yield to his or her temptations. High delayers, Mischel posited, were more socially responsible and less likely than poor delayers to engage in delinquent behavior.[8] High delayers were more mature and autonomous and possessed stronger achievement skills than poor delayers.[9] On the flip side, the inabil-ity to control impulses was correlated with children who lacked "a perma-nent father or father-figure."[10] These were the sort of hypotheses that came decades later to intrigue Goleman — and which Goleman supplemented with his own spin that gratification delay was a skill that could be *acquired*: "There is ample evidence that emotional skills such as impulse control and accurately reading a social situation *can* be learned." Thus, all individuals (not only preschoolers) had the "meta-ability" to resist immediate tempta-tion, but not everyone — yet — had the emotional intelligence required to do so.[11] But they could improve themselves.

In more recent years, the so-called marshmallow test — as read through the lens of Goleman's concept of an emotional intelligence that can be mas-tered with practice and training — fairly swept through the advice-giving culture of the United States (see figure 12). In fact, it would be hard to over-estimate the impact and influence of the marshmallow test on general as-sumptions in the United States about what it takes to lead a happier, more fulfilling and productive existence. The self-help and popular psychology industry found the marshmallow test amenable to a small mountain of mo-tivational agendas — from marriage and relationship manuals to how-to

FIGURE 12. This "Don't Eat the Marshmallow!" coffee
mug represents only one example of how psychologist Walter Mischel's gratification delay experiments
with preschool children have entered the cultural
mainstream; other products include T-shirts and
mouse pads.

investment books. For instance, there were the instructive — if contradictory —
invocations of the marshmallow test, on the one hand, in a guide geared to
helping people exercise better "self-control in an age of excess" and, on the
other hand, in a text on how to "come out ahead in hard times" (by keeping your eye "on the marshmallow!").[12] There were as well countless management and business publications that cited self-control (and often also
Mischel's marshmallow test) in discussions of how and why emotional intelligence represented "the sine qua non of leadership," as Goleman definitively
observed in the *Harvard Business Review* in 1998.[13] A guide on how to lead a
more spiritually meaningful life asked, "Do we have the patience and vision
to achieve what we really want as our end goals or do we lose sight of the

prize because of the big, white marshmallow in front of us?"[14] The marshmallow test turned up in books on how to make a million dollars through clever investments and how to avoid the pitfalls of "short-termism" (because "short-termism causes us to eat the marshmallow").[15] By the time he published *The Marshmallow Test: Mastering Self-Control* in 2014, Mischel himself could not agree more, having fully adopted the framework of emotional intelligence for how to present his own research. But what was being forgotten in all this new enthusiasm for EI and the marshmallow test was the fact that psychological research into the dynamics of self-control had actually had a surprisingly long — and, as it happens, quite politically ambiguous — history.

A close examination of the case study of self-control experiments can contribute an important piece to the larger puzzle of "the disappearance of the social in social psychology."[16] The phenomenon is of course much larger than the work of any one scholar. But what this case may permit us to see with particular clarity is the interplay of internalist and externalist dynamics that help to explain how an inquiry that began by taking a variety of contextual factors into account came over time and gradually to see its "social" context utterly reconfigured. This was due not least to the investigator's own shifting curiosities and motives, as well as to the changing commitments in the disciplinary field, but also to the diverse uses to which policy makers and commentators put the evolving study findings and the changing political climate more generally.

At the latest by 1968, Mischel had gained a controversial reputation by rejecting the prevailing view within personality psychology that "personality traits" remained constant across situations — and thereby could be used predictively. Mischel's work here emphasized the transformative effect of the situation.[17] However ironically, nonetheless tellingly, Mischel's experimental studies on self-control, while initially attentive to contextual factors, over time increasingly became dismissive of those same real-world factors. And yet as they did so, those self-control studies eventually moved from the margins to the center of scientific and popular interest. Mischel's work became more influential the more he left out the world surrounding his subjects; at the same time, his experiments and conclusions were put to work and had real-world consequences. It is not least this paradoxical duality in relation to the world that makes the history of the marshmallow test's evolution and uses not just a fascinating but singular story but rather one representative of far wider trends.

Self-Discipline, Race, and Class

The ideological ambiguity of the marshmallow test, and the ways in which the test became completely interwoven with popular discussions of race (however coded) and class, may be illustrated with the example of an op-ed piece from 2006 (titled "Marshmallows and Public Policy") by *New York Times* columnist David Brooks. Here Brooks observed how crucial it was for think tanks to learn the lessons of the marshmallow test — precisely because the test identified "core psychological traits like delayed gratification skills." Not grasping the import of the marshmallow experiment, or so Brooks lamented, meant that policy experts were "just dancing around with proxy issues" while "not getting to the crux of the problem." Acknowledging the damage done by poverty ("children from poorer homes are more likely to have their lives disrupted by marital breakdown, violence, moving, etc."), Brooks nonetheless derided the idea of "structural reforms" and "structural remedies," noting that their results "are almost always disappointingly modest." Instead he emphasized that "for people without self-control skills, . . . life is a parade of foolish decisions: teen pregnancy, drugs, gambling, truancy and crime," and thus that policy makers should instead be asking "core questions, such as how do we get people to master the sort of self-control that leads to success?"[18]

While such a conclusion may appear — at least superficially — to be free of ideological bias, a review of historical perspectives reveals quite the opposite. For instance, and despite the ways in which Brooks presented his thesis about the failure to delay gratification as self-evidently a road to disaster, this perspective was not in fact the slightest bit self-evident in prior decades. Instead, there was an active debate about the possible correlations between race, class, personality, and life success as well as a debate about the direction of causation.

African American social anthropologist and psychologist Allison Davis, professor of education at the University of Chicago and an early postwar scholar of self-discipline, argued in 1948 *against* a view that class differences were the result of an inability to defer gratification. Davis was unconvinced of the general usefulness of gratification delay. More important, his research showed that socioeconomic status was not a consequence of personality traits. Davis's findings had implications not only for education but also for thinking about the intersections of race and class.

For starters, Davis insisted on the need for scholars and policy makers to grasp that lower-class culture valued entirely different skills and competencies than middle-class culture did. Davis found "pathetic," for instance, the effort to measure lower-class children by middle-class standards of docility and verbal fluency.[19] He called for an educational approach that encouraged children across all classes to be given opportunities for problem solving and that cultivated their abilities to reason, analyze, invent, and imagine. Davis contended that if lower-class individuals did not restrain various impulses it was because they were poor; they were not poor because they were impulsive. Davis focused on how various "systems of control" served to organize Americans along strict class lines, and if there existed "certain gross differences" in behavior between groups, those differences had everything to do with class distinctions. Furthermore, Davis's research empirically demonstrated that there were many more similarities *between* middle-class black and middle-class white families — and many more differences *within* the black community across class divisions. Therefore, Davis hypothesized that all and any differences between whites and blacks "are not related to hereditary, biological, or 'racial' factors." Ultimately, Davis's conclusion was not that the individuals needed to work on themselves but rather that the society needed to change. If educators truly wanted to witness improvements in "the personality development of Negro adolescents," it would become necessary "to reorganize the social and economic structures" of the United States, since only an economic restructuring could conceivably "decrease the social wastage of human lives in our society."[20]

Davis — hardly a lone voice — was part of a much larger set of social scientific conversations about the politics of self-control. In one line of discussion, sociologists, psychologists, and educators in the 1930s and 1940s found that African Americans who sought to adapt to white middle-class standards of self-renunciation were, due to racism, all too often denied the respect and benefits they had expected and hoped to receive. In 1937 psychologist John Dollard wrote that "it can be assumed that human beings never give up possibilities for gratification just for fun; self-restraint is difficult and there must be an adequate social premium on it. The general formula seems to be that the middle-class Negroes sacrifice the direct impulse gains of the lower-class group and expect to have in return the gratifications of prestige and mastery. They *expect* to get them, but the fact is that they are not always paid out according to our cultural model." Like Allison Davis, Dollard too was convinced that efforts on the part of individuals to alter their personal behavior were not enough. "Lower-class culture cannot be changed in a more restrictive direction," Dollard concluded, "until the social situation is changed and the economic and status rewards for labor and self-discipline are increased."[21]

Another line of discussion had to do with class differences among whites. Over and over, investigators established that patterns of child-rearing — including weaning, toilet training, restrictions on movement, and encouragement to take on individual responsibilities — differed dramatically across class lines, with significant effects on the personality structures of the individuals raised in those two divergent worlds. A widely circulated article by psychologist Martha C. Ericson in 1946 (based on one hundred interviews with middle-class and lower-class white mothers in Chicago) found that "middle-class children and lower-class children live in well-differentiated cultures" and that the differences showed up in the resulting personalities; middle-class children were likely to be "subjected to more frustration in learning" and to be "more anxious" than their lower-class counterparts.[22] When educator and psychologist Robert J. Havighurst gave a fuller report on Ericson's findings and incorporated also her research on African American parents, "the results show that the same types of differences exist between middle and lower-class Negroes as between middle and lower-class whites." Like Ericson, Havighurst took the view that the child-rearing practices of middle-class parents caused children to experience "more frustrations of their impulses" than did the practices of working-class parents.[23]

The implicit suggestion that the restrictions on impulses might cause malformations of character in middle-class children found fullest expression in 1946 in Arnold W. Green's "The Middle Class Male Child and Neurosis." Basing his theories on growing up in a small Massachusetts community of Polish immigrant and native-born families, Green noted a relative absence of neurotic symptoms among the working-class immigrant children, in contrast to a concomitant prevalence of emotional problems exhibited by the economically better-situated native-born children. Green leveled his critique at the white middle-class family as a place where the father saw his son as sapping emotional energies and resources better conserved to advance his own status and material progress. And mothers found themselves trapped in a domestic existence where every waking moment was devoted to "the drudgery of housecleaning, diapers, and the preparation of meals." The marriage life (and the marriage bed) languished as a consequence, as the wife watched how "half her working day is spent doing something she does not like, [while] the rest is spent thinking up ways of getting even with her husband." The child caught in this nexus of pain and internalized self-denial soon discovered that "the living room furniture is more important to his mother than his impulse to crawl over it," a fact that "unquestionably finds a place in the background of the etiology of a certain type of neurosis, however absurd it may appear."[24] Green's article was widely reprinted and often cited in the decades to follow. Betty Friedan's *The Feminine Mystique* would rely heavily on Green's analysis to buttress her scathing assessment of the miserable life of the middle-class housewife.[25]

Sociologists through the 1950s and into the 1960s persisted in arguing that when middle-class parents enforced delayed gratification strategies, these produced neuroses in children. Sociologist Albert K. Cohen's *Delinquent Boys*, a pioneering 1955 study in subcultures — however counterintuitively, given its title — perpetuated the *negative* view associated with middle-class attitudes and mores. Such attitudes meant, or so Cohen declared, that a middle-class child was invariably raised to "cultivate the polish, the sophistication, the fluency, the 'good appearance' and the 'personality' so useful in 'selling oneself' and manipulating others in the middle-class world." Cohen scarcely moderated his disdain for middle-class parents who placed an unhealthful premium on their children's "readiness" and "ability to postpone and to subordinate the temptations of immediate satisfactions and self-indulgence in the interest of the achievement of long-run goals." Cohen was overtly sarcastic when he discussed how middle-class parents acted as though "industry and thrift, even divorced from any conscious utilitarian objectives, are admirable in themselves." More often than not, the life of a middle-class child, Cohen added, was "geared to a timetable, to the future as well as the present," and as a consequence the child became "constantly aware of what his parents want him to *be* and to *become*." None of these values, Cohen emphasized, had much to do with love. On the contrary, middle-class children were socialized early to recognize that any love their parents might express toward them was "precarious and contingent": "something to be merited, to be earned by effort and achievement."[26]

Other sociologists who explored the styles and behaviors of delinquent youth in the fifties and early sixties also found the concept of an emotionally warped middle class to be enormously productive. For one thing, and however unexpectedly, this concept enabled sociologists to argue that there were ruptures *and* continuities between "straight" and "delinquent" worlds.

No one wanted to play by the rules *all* the time. It was not only delinquents who had the itch to escape the rat race. *Everyone* sought their "kicks," and so a search for thrills could hardly be labeled "a deviant value, in any full sense," but actually existed "side by side with the values of security, routinization, and the rest." As a significant reflection from 1961 on the phenomenon of delinquency summarized, "In other words, the middle class citizen may seem like a far cry from the delinquent on the prowl for 'thrills,' but they both recognize and share the idea that 'thrills' are worth pursuing."[27]

This bleak assessment of middle-class values and a view that the stricter child-rearing practices of the middle-class family produced children *more* likely to be neurotic did already garner rebukes as early as the first years of the 1950s, and from a variety of angles. Importantly, however, in these critiques nonpermissiveness and insistence on early acquisition of self-control continued to be considered a problem; the question now concerned *which class* was more guilty of strict parenting. One report published in 1954, for instance, came to conclusions that well-to-do families tended in fact to be *more* "permissive" and to "appear to be more tolerant of infantile behavior and to employ less severe punishment in the process of training" than families from more modest class backgrounds.[28] Likewise, a study conducted in 1958 concluded that "the cultural gap" between the child-rearing practices in middle- and working-class families "may be narrowing," in part because Dr. Benjamin Spock's best-selling manual *Baby and Child Care* "has joined the Bible on the working-class shelf" and in part because working-class parents had largely come to accept "middle-class levels of aspiration."[29]

By the early 1960s, sociologist and historian William Sewell—disagreeing strongly with the ascription of unhappy uptightness to the middle class and confirming the newer trend that found middle-class parents actually often to be the more permissive ones—argued that it was in fact difficult to find *any* clear correlation between personality and socioeconomic status. Sewell found fault with Davis's, Havighurst's, and Ericson's conclusions on the restrictiveness of middle-class mothers and the ensuing supposed greater frustration of middle-class children's impulses. But his irritation was directed primarily at what he saw as the overenthusiastic reception of those particular conclusions by the many social science and social work experts who had uncritically absorbed Green's portrait of the neurotic male middle-class child desperate to please his parents. Sewell was articulately dismissive of how a narrowly Freudian fixation on the very early childhood years, and especially on breast and bowels, had gone unchallenged across the social sciences for more than fifteen years. He was most adamant in emphasizing that extant recent studies "offered absolutely no support for the notion that middle-class children more commonly exhibit neurotic personality traits than do children of lower-class origins."[30] And he made a raft of suggestions for how large-scale research on the three-way relationship between socioeconomic status, child-rearing practices, and personality adjustment might be designed more carefully in the future. Missing from Sewell's survey, however, was any attention to the original context in which Davis, Havighurst, and Ericson were writing and the urgent need they had felt to refute racist assumptions about African Americans and thus to disaggregate race from class.

Delinquency and Moral Behavior

In the meantime, a parallel discussion had been taking shape in the interdisciplinary field of criminology that would only later become blended into the debates among psychologists and sociologists about the value (or not) of self-control. In the 1950s, research that correlated impulse control with character development began to have a major impact within criminology. Experts sought scientific methods to predict a predisposition for deviant and criminal behavior. Delinquency was a psychological as well as a social problem, or so criminologists hypothesized, and therefore "the testing of children early and periodically to detect malformations of emotional and character development at a stage when the twig can still be bent" was considered essential for the future well-being of the children themselves and for the greater good of society.[31] If a technique could be found that reliably screened children for delinquent tendencies before those tendencies became full-blown during adolescence, then early preventive steps might be implemented to treat and block those tendencies.

Leading scholars devoted a great deal of time and effort to this line of inquiry. Arguably no single text in this field loomed larger than *Unraveling Juvenile Delinquency*, a major study conducted by Harvard criminologists Sheldon Glueck and Eleanor Glueck and published in 1950. The Gluecks began their research in the late 1930s, whereupon they then devoted a full ten years to a longitudinal analysis of five hundred delinquent and nondelinquent boys. The Gluecks concluded that there were a multitude of factors — temperamental, intellectual, familial, and biological — that contributed to a boy's potential for the development of delinquent behavior later in his life. They hypothesized that the boys who grew up in an environment that was "little controlled and culturally inconsistent" more readily tended to "give expression to their untamed impulses and their self-centered desires by means of various forms of delinquent behavior." One almost invariably found in the future delinquent, the Gluecks noted, that "tendencies towards uninhibited energy-expression" had been "deeply anchored" in the child's "psyche and in the malformations of character during the first few years of life."[32] There were, or so the Gluecks surmised, remarkable continuities between childhood deviance and adult criminality.

The Gluecks additionally designed a Social Prediction Table to be used by social workers, psychologists, educators, and criminologists to forecast a young child's chances of growing from juvenile delinquency into adult criminality. In the course of the 1950s, a number of studies conducted around the United States (as well as Japan and France) tended to validate the usefulness of the Gluecks' Social Prediction Table.[33] By 1960, and building directly on the Gluecks' research, the *New York Times* reported how a longitudinal study of "likely bad boys" conducted in the Bronx had established that "the detection of potential delinquency in children under 6 years old has been reduced to a relatively exact science."[34] Notable as well was the revelatory study *Deviant Children Grown Up*, which further confirmed the Gluecks' analysis. Here sociologist Lee Robins had reviewed discarded records from a child guidance clinic in St. Louis from the 1920s and 1930s and then had tracked down and conducted follow-up interviews with one hundred individuals — now in their forties — who had received counseling at the clinic thirty years earlier. What Robins concluded — as had the Gluecks in the 1950s — was that "adult antisocial behavior virtually requires childhood antisocial behavior."[35] Woven through this thread of research was a general conclusion that exposure to "low moral standards" in the homes where children grew up was an absolutely key factor in predicting future delinquent and criminal behaviors.[36]

Efforts to devise a successful predictive test for childhood deviancy and attempts to dissect the psychological mechanisms of delay and reward were soon to find common cause. This interdisciplinary meeting of methodologies did not, however, happen all at once, perhaps because the history of delay and reward had long involved the use of animals (for example, rats or pigeons) as trial subjects. The efforts to theorize a relationship between a delay in reinforcement and learning response had their origins in the first decades of the twentieth century.[37] These experiments had typically involved either the use of delayed rewards to train animals to perform specific tasks or to predict how variable delays during the acquisition of a skill correlated with greater resistance to extinction of those skills.[38] Once experimental studies turned to the use of children as subjects in the course of the 1950s, these early tests also principally investigated the causal impact of reinforcement delay on learning and skill acquisition.[39] A predictive link between deviancy and reinforcement delay as an area of inquiry arrived somewhat later.

Making this connection between a child's struggle to choose between delayed or immediate gratification and the development of a personality that could (or could not) control deviant behavior was an innovation that can be credited to Walter Mischel. The first delayed gratification experiment Mischel ever conducted involved field research with several dozen "Negro" and "Indian" children between the ages of seven and nine on the West Indies island of Trinidad. Strikingly, Mischel sought to *challenge* the crude generalizations that "the Negroes are impulsive, indulge themselves, settle for next to nothing if they can get it right away, [and] do not work or wait for bigger things in the future but, instead, prefer smaller gains immediately," while "the Indian is said to deprive himself and to be willing and able to postpone immediate gain and pleasure for the sake of obtaining greater rewards and returns in the future." (Mischel presented these generalizations as each group's own expressed prejudices about the other.) In his experiment, Mischel *did* find some correlation between ethnic group and choice behavior (with more "Negro" than "Indian" children seeking immediate gratification). But he also found evidence that "the presence or absence of the father" in the home mattered just as much — regardless of the child's racial background. And he likewise found that older children were better able to delay — again, regardless of race. This suggested that an ability to resist immediate temptation represented "a learned behavior which is, in part, a function of the expectancy that the promised reinforcement will issue from the social agent in spite of time delay."[40] In short, at this point Mischel's analysis included attention to the question of how reliably the child could count on the testing adult to come through with that larger piece of candy. The trustworthiness of the tester was here taken into account.

Mischel's subsequent decision to moralize about the meanings of delayed gratification became evident little by little over a period of some years. Relevant not least was Mischel's graduate work with Ohio State clinical psychologist Julian Rotter, whose own interest in learned behavior had led him to develop the theory of "locus of control" (as discussed in chapter 1). Rotter had written already in 1954, in *Social Learning and Clinical Psychology*, that for a child to "discover that some current satisfactions for maladaptive behavior are likely not to continue in the future," that child must learn that "the goals he sees as providing satisfaction should be ones that it is possible for him to attain, and as much as possible they should be goals that are not likely to be followed by strong punishments, either delayed or immediate."[41]

But by the mid-1960s Mischel had begun to delve ever more deeply into the history of a "character education" movement whose heyday had been in the early decades of the twentieth century. Mischel became particularly interested by the Character Education Inquiry conducted at Teachers College, Columbia University, during the 1920s. This massive social scientific project had been based on elaborate experiments in which more than ten thousand children in nearly two dozen communities across the United States were provided opportunities where they could either engage in honest conduct or participate in cheating, lying, or stealing. The project's objective was to identify possible ways in which educational programs could be useful for inhibiting deceitful behaviors. However, the researchers' results were wholly discouraging: "The mere urging of honest behavior by teachers or the discussion of standards and ideals of honesty, no matter how much such general ideals may be 'emotionalized,' has no necessary relation to conduct."[42] In reviewing these conclusions, the *Journal of Education* in 1931 dismally determined, "What children are learning of self-control, service, and honesty seems to be largely a matter of accident. There is little evidence that they are being influenced by effectively organized moral education."[43] Yet a full generation later, Walter Mischel drew different conclusions from the "character education" movement as he connected his own research on delayed gratification with the study of what he designated "moral behavior," which he defined as the ways in which individuals demonstrated a "capacity to regulate, judge, and monitor their own behavior even in the absence of external constraints and authorities."[44]

In 1964 Mischel made explicit his hypothesis that there was a correlation between social deviance and incapacity to delay gratification. Mischel began to position his research into gratification delay as being in direct lineage with studies that had "investigated resistance to temptation, and tried to correlate the subject's reactions when under pressure to violate his standards with indices of 'guilt' and with aspects of parental disciplinary techniques."[45] In one of the few experiments Mischel conducted that utilized the strategy of deception of test subjects (a study undertaken collaboratively with Carol Gilligan, then a doctoral student at Harvard), teenagers kept their own scores while they played an arcade-style game. However, the points they were able to earn were rigged so that "the scores they got made it impossible to win a badge: to get a badge, they had to cheat by falsifying their scores, and to win a better badge they had to fake it even more."[46] Although this test had concerned frustrating circumstances in which actual skill was never appropriately rewarded, rather than a test for deferring a gratification that could be trusted eventually to arrive, Mischel and Gilligan extrapolated more broadly: "If the subject is to resist the temptation and to refrain from deviant behavior, he must be able to defer immediate gratification."[47] A new experimental psychological paradigm that emphasized the responsible need for self-control was born.

Poverty, Self-Indulgence, and Public Policy

The Negro Family: The Case for National Action, written by political scientist (and assistant secretary of labor) Daniel Patrick Moynihan and published by the U.S. Department of Labor's Office of Policy Planning and Research in 1965, likely still remains best remembered — and reviled — for making an argument that the African American community suffered from a "tangle of pathology." Chief among these pathologies, or so Moynihan held, was that the black family "has been forced into a matriarchal structure which, because it is so out of line with the rest of the American society, seriously retards the progress of the group as a whole."[48] Intended to call attention to the long-term damages caused by white racism, the Moynihan Report achieved an opposite effect, quickly becoming "the focal point for one of the more vociferous public debates about policy in the history of the nation," while Moynihan himself "was pilloried as a racist and the foremost neoconservative on matters of race."[49] The backlash against the Moynihan Report was sustained and intensely hostile.[50] As historian Ellen Herman observed, civil rights activists and their supporters assailed Moynihan for implying that the problems that afflicted the African American family "were primarily personal and psychological" instead of acknowledging how "racial oppression produced social pathology rather than vice versa." Thus, or so his critics asserted, what Moynihan suggested was that "institutional racism and discrimination could be deemphasized or even eliminated as a terrain of governmental action."[51] The Moynihan Report represented, psychologist and political activist William Ryan famously declared in 1971, a classic case of "blaming the victim."[52] This characterization has largely stuck to the Moynihan Report for more than half a century.[53]

At the same time, the Moynihan Report brought little that was new to light in its explanations for the root causes of African American poverty. Rather, what the Moynihan Report did was cull analyses from several well-regarded expert texts in the behavioral and social sciences. For instance, its discussions of black family life made repeated references to *A Profile of the Negro American* by social psychologist Thomas Pettigrew.[54] And psychological research informed the report in at least one additional and important respect.

This had to do with the psychological roots of juvenile crime and delinquency. Citing several of Mischel's early experiments, the Moynihan Report accepted as simple fact that an individual's ability to defer gratification correlated with that individual's family structure. Children in "fatherless homes" sought out "immediate gratification of their desires," and those youth (in their "hunger for immediate gratification") were, in turn, "more prone to delinquency, along with other less social behavior." Turning back to the work of Pettigrew, the report added that boys from "high delinquency neighborhoods" raised in "stable, intact families" were able to resist — and rise above — the misfortune of their surroundings.[55] Although it has been largely unnoticed in the voluminous scholarship on the Moynihan Report, the report thus promoted a hypothesis that gratification delay enabled persons to advance socioeconomically, while self-indulgence condemned persons to a life replete with personal problems and squandered opportunities.[56]

A view that an inability to delay gratification might be correlated specifically to African American poverty persisted in the public policy literature of the late 1960s and early 1970s. In 1968 the *Manpower Report of the President* invoked a prevalent view that "explanations of the job behavior of low-income Negroes and others who have difficulty in getting and keeping jobs" included such "social-psychological factors" as "attitudes, aspirations, motivation (especially achievement motivation), ability or willingness to defer gratification."[57] That same year, political scientist Edward Banfield's *The Unheavenly City* was saturated with (racially coded) assumptions about the inability of poor people to postpone gratification. For instance, Banfield observed how "the lower-class individual lives from moment to moment. . . . He is therefore radically improvident: whatever he cannot consume immediately he considers valueless. His bodily needs (especially for sex) and his taste for 'action' take precedence over everything else — and certainly over any

HUMAN BEINGS ARE NOT VERY EASY TO CHANGE AFTER ALL

An unjoyful message and its implications for social programs

A while back there was a severe shortage of electricity in New York City, and Columbia University tried to help out in two ways: A card reading "Save a watt" was placed on everyone's desk, and janitors removed some light bulbs from university corridors. The ways in which this shortage was made up for illustrate two major approaches to social problem solving. One approach is based on the assumption that people can be taught to change their habits, that they can learn to remember to switch off unused lights. The second approach assumes that people need not, or will not, change and instead alter their environment so that, even if they leave light switches on, watts are saved.

The prevalent approach in the treatment of our numerous and still-multiplying social problems is the first. Imbedded in the programs of the federal, state, and city governments and embraced almost instinctively by many citizens, especially liberal ones, is the assumption that, if you go out there and get the message across—persuade,

BY AMITAI ETZIONI

$27 million is used to make nonsmokers out of smokers—that is, to try to change a basic habit—no significant effect is to be expected. Advertising molds or teases our appetites, but it doesn't change basic tastes, values, or preferences. Try to advertise desegregation to racists, world government to chauvinists, temperance to alcoholics, or—as we still do at the cost of $16 million a year—drug abstention to addicts, and see how far you get.

In fact, the mass media in general have proved to be ineffectual as tools for profoundly converting people. Studies have shown that persons are more likely to heed spouses, relatives, friends, and "opinion leaders" than broadcasted or printed words when it comes to deep concerns.

Another area in which efforts to re-make people have proved glaringly inefficient is that of the rehabilitation of criminals. We rely heavily on re-educational programs for prisoners. But it is a matter of record that out of every two inmates released, one will be rearrested and returned to prison

FIGURE 13. The opening section of "Human Beings Are Not Very Easy to Change after All," a *Saturday Review* article from June 3, 1972, written by Amitai Etzioni and illustrated by Charles Slackman. Reprinted by permission of Amitai Etzioni and Betteanne Terrell Slackman.

work routine."[58] Such were the terms of debate by 1972 that Amitai Etzioni, director of the Center for Policy Research at Columbia University and writing in the *Saturday Review*, could suggest (in language that could also be read as racially coded) that federally sponsored training programs that sought "to try to change" the "work behavior" of "disadvantaged" individuals were destined to fail because "human beings are not very easy to change after all" and that training programs would fare far better if they helped such persons "choose jobs compatible with their personalities"[59] (see figure 13). A tautological argument emerged; people who lacked an ability to resist immediate satisfactions were more likely to fail in socioeconomic terms, while lower-class persons suffered from an inability to defer gratification. It was, as historian Alice O'Connor has noted, a "profile of lower-class personality

disorders" that was "contradictory, culturally biased, and remarkably sim-
plistic at times" and yet continued to gain ground — at least among illiberal
social policy researchers.[60]

Some social scientists remained unconvinced. A widely cited article already in 1965 argued, for instance, that the ascendant "poverty and self-
indulgence" paradigm that used a "deferred gratification pattern" (DGP) to
explicate "'lower class behavior'" was just "a thin reed on which to hang analyses of behavior," and that therefore DGP could not be "mechanically applied
to interpret low income life." The critique discussed previously unpublished
psychological research that tested the DGP; an experimenter promised candy
to two randomly assigned groups of children from various racial and social
backgrounds, but he then broke his promise to one of the two groups. When
he repeated the experiment, it was not class or race that determined a child's
ability or inability to defer gratification but rather whether or not the child
belonged to the group that had been deceived. *"The situational variable, then,
rather than class affiliation determined the ability to delay,"* this research
found. The article concluded that "at the level of social science analysis, the
verdict on the DGP is 'not proven.'"[61] And yet a commonsense view that jum-
bled causation and correlation and was drenched in associations with both
class status and racial identity — and that assumed that a person's inability to
control his or her immediate impulses predicted that person's ability to succeed in life — would simply not go away.

Crime and Human Nature

In 1971, Mischel directly suggested that there was a link between an "inability to tolerate delay of reinforcement" and "the development of criminal
behavior." Mischel observed that "surprisingly, in spite of the important role
of delay of reward in many personality theories," there still existed "little
systematic research"— beyond his own — on establishing this link between
criminality and a person's lack of self-control.[62] By 1974, Mischel started to
make more pronounced arguments about causation and not just correlation between class status and a demonstrated capacity for impulse control.
He noted that the "middle and upper (in contrast to lower) socioeconomic
classes" exhibited "higher intelligence, more mature cognitive development, and a greater capacity for sustained attention" and that "membership
in the lower socioeconomic classes" correlated with lower levels of these
same qualities. But he also began to itemize "the magnitude of the social

problems" associated with "inadequate voluntary delay and deficiencies in self-control." Despite his qualifying use of the word "partial," Mischel contended that "inadequate delay patterns" often were "the partial causes of antisocial and criminal behavior, violence and physical aggression, and failure to achieve reasonable work and interpersonal satisfactions." In sum, Mischel made what was at that point the most far-reaching case to date that "deficiencies in voluntary delay may become a major source of frustration by guaranteeing the individual an endless chain of failure in our culture." Mischel here effectively did not shy from making quite grandiose claims for his marshmallow-derived findings, as now the fate of society itself was said to be at stake: "Basic to most philosophical concepts of 'will power' and the parallel psychological construct of 'ego strength' is the ability to postpone immediate gratification for the sake of future consequences, to impose delays of reward on oneself, and to tolerate such self-initiated frustration. It is difficult to conceive of socialization (or, indeed, of civilization) without such self-imposed delays."[63] Although almost certainly unintended by Mischel, these maneuvers soon opened up his research to a whole new range of ideological applications.

It was Harvard psychologist Richard J. Herrnstein who first imported a version of Mischel's concepts about self-imposed delays, socioeconomic class status, and "deviant" behavior into the discipline of criminology. In the late 1950s and early 1960s, Herrnstein had published animal research (with pigeons) on the use of variable-interval time-delayed reinforcements.[64] He had clearly followed the progress of Mischel's research. This would be confirmed in 1985 when, in *Crime and Human Nature*, a tome Herrnstein cowrote with political scientist James Q. Wilson, the authors not only offered favorable assessments of the longitudinal studies of Sheldon and Eleanor Glueck as well as sociologist Lee Robins but additionally made honorable mention of Mischel's experimental research on delayed gratification. With an overt reference to Mischel, the authors noted how "offenders are much more likely than nonoffenders to prefer an immediate small reward to a delayed larger one." And citing the same 1974 passage from Mischel quoted above, Herrnstein and Wilson portentously concluded how the "predisposition to impulsiveness" in children, "when combined with a family setting in which rewards and penalties are not systematically made contingent on behavior," in turn "leads to inadequate socialization" and places "a strain on civilization."[65] Mischel's impact on the text's analysis of what kind of person commits criminal acts was unmistakable — if also selectively skewed.

The idea that familial breakdown led to deficiencies in self-control that in turn produced criminal behavior was not, presumably, the precise viewpoint Mischel meant to promote, but his ideas proved very usable for an unabashedly right-wing causational theory.

There were equally tendentious efforts to cement connections between the (in)capacity to delay gratification and a host of behavioral and cognitive issues. Mischel had initially proposed in 1962 that high delayers were on the whole more intelligent than poor delayers.[66] But it was only in the course of the 1980s that a full-blown "cognitive deficit" theory linking low IQ and weak impulse control to criminal behavior came to popular and scholarly fruition. Strikingly, here again the pronouncements of psychologist Herrnstein proved instrumental.[67] But the grand leap from an inability to resist temptation to low IQ and a propensity to commit crimes made its boldest entrance only in 1990. "Criminal acts provide *immediate* gratification of desires," Michael R. Gottfredson and Travis Hirschi wrote that year in their milestone book, *A General Theory of Crime*, adding, "The cognitive requirements for most crimes are minimal." For Gottfredson and Hirschi, poor impulse control explained "all crime, at all times, and, for that matter, many forms of behavior that are not sanctioned by the state." For these authors, low self-control was "for all intents and purposes, *the* individual-level cause of crime," and furthermore "the search for personality correlates of crime other than self-control" was "unlikely to bear fruit."[68] That blanket declarations like these elicited fierce response would be an understatement; *A General Theory of Crime* would be the most widely discussed (and critiqued) text in the field of criminology published in the last generation.[69] Nonetheless, despite vigorous criticism, the general thrust of the analysis had remarkable holding power.

How Children Succeed

If poor people possessed low impulse control and lower IQs and were more prone to criminal behavior, what might be asserted about the well-to-do? Was there no criminal malfeasance at the higher echelons? Richard Herrnstein had argued at least since 1971 that U.S. society was developing into a meritocracy with high IQ scorers dominating the upper class (and low IQ scorers slipping into the socioeconomic bottom rung).[70] However, a position that the wealthy *deserved* their cozy status because they possessed the smartest genes truly came to prominence only with the publication of *The*

Bell Curve. Here Herrnstein and political scientist Charles Murray popu-
larized the concept of "cognitive stratification," arguing that IQ scores essentially determined individuals' economic fortunes in life. Why would a
"cognitive elite" not float "toward the upper income brackets," Herrnstein
and Murray pondered, as well as come to dominate "most of the institutions
in society" while also acquiring "some of the characteristics of a caste"?[71]
Nor would Murray (after Herrnstein's death the same month in 1994 when
The Bell Curve was published) ever be prompted to revise himself. On the
contrary, Murray in 2012 observed how "cognitive segregation" was "bound
to start developing" in the United States "as soon as unusually smart people began to have the opportunity to hang out with other unusually smart
people." Brainy (and successful) parents quite simply produced brainy (and
successful) children. This was so, Murray declared, because brainy parents
possessed both excellent foresight and self-discipline.[72] The well-to-do were
experts at "intense planning" because they were bright, and well-to-do children were bright because their parents had been expert planners. Disability
studies scholar Michael Bérubé already in the 1990s characterized Murray's
vision of the world as nothing other than social Darwinism.[73] And yet as
problematic as Murray's familial fantasies were, they nonetheless continued
to accrue ideological weight and intellectual gravitas.[74]

There must have been extraordinary hunger for an alternative framework.
Against the position of right-leaning social scientists that human nature was
essentially intractable (and therefore the liberal policy programs associated
with the 1960s and early 1970s had been doomed to failure), a new theory
began to be formulated at the turn of the millennium: a theory of human
nature that people *could* change — though *by themselves* and therefore *without* the crutch of a government-sponsored social welfare system. The notion
of tractability — which had been associated with the "social engineering"
of liberal policy programs in the 1960s — had now itself shifted into a more
politically muted (and less social and structurally oriented) psychological
analysis. Theorists of self-control who counted themselves opposed to the
right-wing perspective that personality was set in stone from birth (and was
therefore genetic or in some other way biological) tended to operate on the
premise that they had never met a self-regulating behavior they did not like.

Gone were the post–New Deal era theories that too much delay of gratification warped middle-class children, as adults and children alike were now
being encouraged to master their impulses. But gone too was the conviction that IQ outweighed everything else. Instead there were the new and

immensely influential ideas of University of Pennsylvania developmental psychologist Angela Duckworth. Duckworth had worked repeatedly with Mischel and turned what she learned about gratification delay into self-control exercises to help schoolchildren improve their academic performance. She further developed what she called a "grit scale" to measure perseverance and motivation. Duckworth's philosophy was as simple as had been Mischel's initial marshmallow experiments. "There may be no such thing as 'too much' self-control," she said in 2011.[75] But were race and class really gone?

An education reform movement took off that posited the value of cultivating the impulse control of children of all colors and socioeconomic backgrounds. Paul Tough's best-selling *How Children Succeed: Grit, Curiosity, and the Hidden Power of Character*, published in 2012, became the movement's manifesto, as it carefully linked psychological theories about self-control to a novel "cognitive hypothesis" buttressed by neuroscience that "what matters most in a child's development" is "not how much information we can stuff into her brain"; rather, "we are able to help her develop a very different set of qualities, a list that includes persistence, self-control, curiosity, conscientiousness, grit, and self-confidence."[76] Tough relied extensively on Duckworth's research when he offered these conclusions.

Despite recurrent invocations of Mischel as a key authority, though, these new perspectives concerning the virtues of self-discipline did not have their origins solely in Mischel's work on delayed gratification. Another strand of thinking from the burgeoning field known as positive psychology blended into the work of the newest educational reformers. As cognitive and behavioral psychologist Martin E. P. Seligman wrote in 1991 in *Learned Optimism*, a foundational text of the positive psychology movement and a key counterpoint to his prior work on "learned helplessness," his view represented a rejection of the classic behavioral experiment involving reinforcements and rats.[77] Give a rat a pellet every time he pressed a bar, and he will learn to press the bar to receive a pellet. Never give a rat a pellet every time he pressed a bar, and he will stop pressing the bar to receive a pellet. Give a rat a pellet only once in a while when he presses a bar, and he will continue to press the bar for a long while before complete extinction of this impulse takes over — and he quits pressing the bar. Seligman argued that this experiment with rats did *not* work with humans. Seligman concluded instead that human behavior was controlled "not just by the 'schedule of reinforcement' in the environment but by an internal mental state, the explanations people make for why the environment has scheduled their reinforcements in this way." In terms of people's persistence, then, "what really mattered" was "the way people think about the causes of successes and failures," not so much whether they actually succeeded or failed.[78] Rather than being predominantly focused on developing cognitive skills (like IQ), positive psychology directed attention toward "building and using your signature strengths" because "the good life is using your signature strengths every day to produce authentic happiness and abundant gratification," as Seligman wrote in 2002 in *Authentic Happiness: Using the New Positive Psychology to Realize Your Potential for Lasting Fulfillment.*[79]

As for the single most important predictor of personal achievement, Seligman and Duckworth together wrote in 2005 (based on a longitudinal study of more than one hundred eighth-graders) that it had virtually everything to do with "self-discipline": "We found that self-discipline predicted academic performance more robustly than did IQ. Self-discipline also predicted which students would improve their grades over the course of the school year, whereas IQ did not."[80] In 2013 Duckworth was awarded the prestigious MacArthur Foundation grant for her research on the critical role that self-control played in educational achievement. Seligman and Duckworth's non-IQ approach impressively avoided the elitism (and given the mutual imbrication of the class and racial stratifications in U.S. society, the not-so-hidden racism) of Murray and Herrnstein. But that did not mean that there were not class- and race-based assumptions about the direction of causation between individual character development and life success built into the studies on which they grounded their theories.

Controlling Ourselves

In 2011 a major longitudinal study concluded that an inability in early childhood to self-regulate impulses represented a significant public policy concern. Tracking the lives of a thousand children over a thirty-year period, the study determined that a child's early failure to develop the skill of self-control strongly correlated with a mass of unfortunate adult outcomes — from more difficulties managing personal finances to more problems with drugs and alcohol to higher chances of having engaged in criminal behavior. "Given that self-control is malleable," the study noted, "it could be a prevention target, and the key policy question becomes when to intervene to achieve the best cost-benefit ratio, in childhood or in adolescence?"[81]

There can be little doubt that the lessons derived from the marshmallow test have greatly accrued cultural and political capital in recent years. This has to do with at least three (perhaps interrelated) reasons. The first is a dramatic increase in scholarly and popular attention to self-regulation in childhood. The second is the new century's enjoinments not just to children but also to adults to struggle over their own impulse control issues. This puts the preoccupation with self-control squarely under the rubric of incessant U.S. concern with self-improvement and self-optimization. Moreover, much of this line of discussion has been infused with references to new research from the neurosciences, as increasing numbers of social cognitive neuro-scientists have announced their intention to utilize brain scans to identify the neural correlates of self-control and gratification delay. These studies often link their conclusions about neural activity and the capacity to resist temptation directly to the Mischel experiments; a striking example is an-other study from 2011 that gathered together a group of the original Stanford preschoolers (now in their mid-forties) so that they might undergo neuroim-aging to test their "behavioral and neural correlates of delay of gratification 40 years later."[82] As a further study from 2013 summarized, the search was definitely on for "a biological marker of self-control ability."[83]

A third reason for the growing relevance of the marshmallow experiments and the obsession with the newly cherished ideal of self-control, however, has to do with the long, unwinding backlash against "the 1960s"— or at least everything that the decade is said to have stood for (that is, self-indulgence, hedonism, narcissism, and a complete lack of personal restraint). That there had been a major time lag between when Mischel conducted his initial ex-periments on the delay of gratification and when they finally became wildly popular cannot be a complete coincidence. Mischel was publishing his re-sults on self-control all through the 1960s, 1970s, and 1980s, but it was only in the 1990s that the phenomenon of gratification delay among children began to take on the intense cultural and political power that it subsequently was understood to possess. Before the 1990s, social and personality psychology textbooks tended scarcely to acknowledge the marshmallow study.[84] Nor did proponents of the social cognitive revolution in the 1980s demonstrate much interest in these studies; a good example is Susan Fiske and Shelley Taylor's *Social Cognition* of 1984, which thematized Mischel's theories on personality and assessment but made no mention of his research on the "cognitive mech-anisms" involved in the delay of gratification.[85] Into the 1990s, gratification delay was still a subject largely unworthy of citation. The most influential texts before the publication of Goleman's *Emotional Intelligence* to bring up

Mischel's work on self-control did so for reasons all their own; these texts included psychologist Albert Bandura's *Aggression: A Social Learning Analysis* (1973) and Carroll E. Izard's *Human Emotions* (1977). In other words, the induction of the marshmallow test into the canon of truly classic and essential psychological experiments took a very long time. And this shift had a good deal to do with a more general turn away from a concern with structural conditions — in the field of experimental psychology and in U.S. media discussion as a whole.

In more recent years, major media outlets have eagerly promoted the neuroscience of gratification delay. A cover story for *Newsweek* magazine in 2011 on "the new science behind your spending addiction" profiled Mischel's marshmallow experiments when it discussed research into "the brain's saving and spending circuits." *Newsweek* noted how several fMRI studies had identified heightened activity in high delayers' "thoughtful, rational prefrontal cortex" as well as their "inferior frontal gyrus which inhibits the 'I want it now' impulse." By contrast, low delayers had reduced activity in these advanced brain regions but more activity within the (evolutionarily more primitive) limbic system. Yet *Newsweek* reported as well that there was hope for individuals who struggled to resist their impulses. For one thing, there was preliminary neuroscience research at Columbia University and elsewhere into the development of a "noninvasive 'zapping' technology, called transcranial magnetic stimulation (TMS)" that might be able to "rev up" the prefrontal cortex — and thus artificially block the cravings produced by the "I want it now" circuitry. Other techniques mentioned by *Newsweek* for instilling "a talent for saving" involved injecting low delayers with "a squirt of the hormone oxytocin," since oxytocin had in some tests been demonstrated not only to reduce anxieties but also to boost self-discipline.[86] An article in *Time* magazine from 2012 on the neuroscience of self-control opened, "Pity your prefrontal cortex — the CEO and chief justice of the bedlam that is your brain." The message (again) was that the prefrontal cortex had the thankless task of corralling a person's "most decadent appetites — for drinking, gambling, eating, smoking, shopping, sloth, sex." In a confused effort to mix neuroscience with theories of evolution, *Time* conceded that while there might be "no survival value" in the act of shopping or gambling (though presumably in sex), these activities provided "kicks of their own" that enabled them — as neuroimaging studies had begun to show — to "sidestep evolution and pick the chemical locks of the brain's pleasure centers directly."[87] Once again the focus skirted social conditions and pervasive cultural inducements to indulge in consumption and pursue pleasures and instead placed an emphasis on what went on *inside* the brain — where the real battle over the lures of instant gratification played out.

Rational Snacking

In the course of the half century or more in which psychologists have studied self-control and delay of gratification in children, a number of themes recur. One has to do with the puzzle of the relationship between *correlations* (of socioeconomic status, race, father-absence, and the like) and *causation*; closely related is the perpetual riddle of the *direction* of causation. Does poverty cause low self-control, or does low self-control lead to poverty? Also important is the *predictive* value of a particular identified phenomenon. Does a character trait demonstrated in childhood have a welter of later consequences? And how valid are findings extrapolated from laboratory settings, anyway? No less important is the question of *valuation*: Is a particular skill or trait good for you or not? And not least there is the issue of the potential *changeability* of human nature. Is character something you can work on? Does your attitude about your situation matter as much as or more than the situation itself? What are the best methods for getting at and sorting through all these issues? Each of these problems has been raised by scholars concerned about ongoing trends in the field.

Skeptics have consistently expressed doubts about the predictive value of laboratory-based testing of gratification delay. Already in the late 1980s, psychologist Jack Block noted that results derived in "the ordinary situations of life in which adaptive delay is so important typically involve both immediate and delayed rewards that are vastly larger, more powerfully motivating, and more conflicting than anything that can be ethically administered in a brief, experimental situation."[88] But Block was also concerned about the *value* of impulse control. In 1996 he argued that while "compliance with parental prescriptions" or "reflexive, unthinking deference to internalized proscriptions" was "developmentally advancing for the child when it occurs . . . such inhibition or compliance does not necessarily represent an adaptively desirable endpoint." And he observed that "adaptability in the long term requires more than the replacement of unbridled impulsivity, or undercontrol, with categorical, pervasive, rigid impulse control."[89] Other psychological studies attested to the unhealthful consequences (for example, a lack of spontaneity, a risk for depression, a tendency to binge or engage in reckless behaviors) that can be associated with the persistent struggle to attain unrealistic objectives.[90] Here, then, there were echoes of older worries about the psychological downsides of too great a preoccupation with upward mobility. A psychological researcher told the *New York Times* in 2008, for instance, that it could be crucial to recognize that a "relentless pursuit of goals" came with "a cost" and was potentially detrimental to a person's well-being.[91] Despite these caveats, as education theorist Alfie Kohn wrote also in 2008, an argument that self-discipline was "required for achievement" and that "its absence is seen as a sign of self-indulgence and therefore of moral weakness" represented a viewpoint that was no longer "limited to talk radio or speeches at the Republican convention" but had now become "threaded through the work of key researchers who not only study self-discipline but vigorously insist on its importance."[92] (Angela Duckworth was quick to counter Kohn's criticism. "As we (and most psychologists) use the term," she contended, "self-discipline is not the ability to accomplish goals which others deem desirable. Rather, self-discipline is the ability to marshal willpower to accomplish goals and uphold standards that one personally regards as desirable.")[93]

In the meantime, in 2013, a doctoral candidate in brain and cognitive sciences at the University of Rochester, Celeste Kidd, sought to challenge the entire conceptual edifice on which the marshmallow experiment was based. Was it really accurate to conclude that the marshmallow test revealed anything intrinsic about a child's long-term capability to negotiate all manner of challenges in life? Or was the marshmallow experiment simply a test of whether an individual child *trusted* that the environment was reliable? Kidd, who had worked as a volunteer in a shelter for homeless families before she attended graduate school, had her own objections to how the answers to these questions were assumed to be self-evident. "Delaying gratification is only the rational choice if the child believes a second marshmallow is likely to be delivered," she told *Scientific American Mind.*[94] Mischel himself had, after all, considered it important (except in the one arcade-game study of cheating) that his child subjects learned to trust *him*. And the 1965 critique of the "deferred gratification pattern" (that had emphasized how the race and class of the children was irrelevant) observed that what mattered in a child's decision-making was whether he or she belonged to a test group that had been lied to by the tester. But Kidd was after something far broader. However inadvertently, then, Kidd placed herself in a lineage with Allison Davis, John Dollard, and others, who had, after all, suggested much the same thing since the 1940s. In her study "Rational Snacking," Kidd wrote that delay of gratification experiments did not test "self-control" so much as they tested for "beliefs about the stability of the world." She concluded that for a child who had not experienced reliable rewards in life, "the only guaranteed treats are the ones you have already swallowed." In other words, children who chose immediately to gratify their needs might well have been "strongly influenced by rational decision-making processes."[95] Or as the title of an article about Kidd's research succinctly put it, "Delaying Gratification Is about Worldview as Much as Willpower."[96]

Conclusion

The history of IQ was always a history of social ideology, historian Paula Fass wrote already in 1980, while she observed as well that the IQ test was initially meant to serve *progressive* ends. IQ, as it was developed in the first three decades of the twentieth century, introduced a metric for "what seemed to be the fairest and the most practical, as well as the most culturally felicitous, organizing principle" to measure an individual's talent in an era of social upheaval and rapid cultural change — even as it quickly came to be used to justify the maintenance of class and racial hierarchies.[97] This chapter has suggested that the history of emotional intelligence is also a history of social ideology and that self-control —"the master aptitude" that Goleman most associated with EI — has been an ambiguous concept intended to resolve dilemmas raised by the racialized assumptions that had recurrently tainted the reputation of IQ testing.[98] While IQ was said to be innate, EI was argued to be the opposite; EI could be acquired, its proponents argued, and it could be taught. And self-control was the principal marker of having achieved greater EI. In an article on EI in 1995, *Time* magazine made the connections clear. It not only opened its cover story on EI with a laudatory summary of the marshmallow test ("It seems that the ability to delay gratification is a master skill, a triumph of the reasoning brain over the impulsive one") but went on uncritically to applaud the development of "emotional literacy" programs that were "designed to help children learn to manage anger, frustration, [and] loneliness" as educators sought actively to reevaluate "the weight they have been giving to traditional lessons and standardized tests."[99] EI (in the first decades of the twenty-first century), like IQ (in the first decades of the twentieth century), became a concept used to evaluate individuals in a fashion that (its advocates argued) was equitable and unbiased. That IQ was perceived to be a flawed metric only caused another metric to be invented.

But when EI arrived on the scene, it did so with its own complicated and historically heavy political baggage, one ineluctably linked to the shifting meanings associated with self-control.

The very concept of impulse control had once been a profoundly racialized one. The way that implicit now-you-see-it-now-you-don't references to race have moved in and out of focus in the literature on impulse control reveals much about the ideological work done by any experimental theory that diverts concern from social and political contexts and toward the individual's success or failure at self-management. But the concept of impulse control was all along very much also a class-tied concept. Just as direct references to race largely disappeared by the 1990s, so too a second great erasure by the early twenty-first century removed explicit acknowledgment of socioeconomic status from the burgeoning literature on self-improvement. The seemingly universal address of the current injunctions to all individuals to upgrade their self-control skills now masked how these ubiquitously repeated injunctions directed attention away from the open secret of (ever more rapidly widening) class divisions — and the descent of a (formerly more secure) middle class into greater and greater insecurity.[100] The study of gratification delay, then, cannot be read apart from a political history of postwar America, even though so many contemporary promoters of the value of impulse control have treated it as a transhistorically valid ideal. But among the many lessons that might be drawn from the history of the study of gratification delay is that trends were under way to leave the real world and its vicissitudes firmly outside the experimental laboratory already long before new brain imaging technology made the exclusion of that real world a practical necessity.

FIVE

Neuroscience, Race, and Intelligence after *The Bell Curve*

LTHOUGH IT MAY NOT mainly be remembered for this reason, Hillary Clinton's best-selling book of 1996, *It Takes a Village*, devoted considerable space to a discussion of brain science and early child development. Clinton wrote of "recent discoveries in neuroscience, molecular biology, and psychology"—discoveries that have "given researchers a whole new understanding of when and how the human brain develops." She wrote of research on the amygdala, the region of the brain where emotions resided. She wrote of psychologist Daniel Goleman and the importance of his theories on emotional intelligence. Principally, however, Clinton focused on the new science surrounding brain development in early childhood. "Early experience—especially how infants are held, touched, fed, spoken to, and gazed at—seems to be key in laying down the brain's mechanisms that will govern feelings and behavior," Clinton observed. She added that "a child's character and potential are not already determined at birth." Time and again, Clinton cautioned how the child's "first few years" were crucial ones because "by the time most children begin preschool, the architecture of the brain has essentially been constructed."[1]

Clinton's discourse on how the brain develops in early childhood represented her critical riposte to Richard Herrnstein and Charles Murray's *The Bell Curve*. Clinton wrote that Herrnstein and Murray had simply been wrong to imply—as she summarized them—that "if nothing can alter intellectual potential, nothing needs to be offered to those who begin life with fewer resources or in less favorable environments." Clinton noted that this erroneous position went hand in hand in *The Bell Curve* with the book's other "politically convenient" ideas, including that "intelligence is fixed at birth" and that intelligence is "part of our genetic makeup that is invulnerable to change." She cited evidence from a study that witnessed children who were "at risk" thriving after they had been "exposed at an early age to stimulating environments." She paused to discuss at some length psychologist Craig Ramey's Abecedarian Project in North Carolina; Ramey had provided dozens of mainly African American and poor infants with both early educational programs and excellent nutrition only to see these children's IQs improve dramatically (gains that did *not* fade over time). Clinton wrote that *The Bell Curve*, therefore, was "not only unscientific but insidious." And she concluded, "If we as a village decide not to help families develop their children's brains, then at least let us admit that we are acting not on the evidence but according to a different agenda."[2] This was scarcely a move by Clinton away from a concern with either public policy or politics and into the dense underbrush of neuroscience. On the contrary, this was an unambiguous attempt by Clinton to take compelling new findings from brain science and put them to work *as* politics and public policy.

In April 1997 President Bill Clinton and the First Lady cohosted a conference on "what new research on the brain tells us about our youngest children." Experts from several fields were invited to speak, including neuroscientist Carla Shatz, who offered that a child's brain development resembled "the problem of stringing telephone wires from one city to another." Shatz added that "the problem of wiring isn't over" once the main "trunk lines" are laid down. There remained "a second phase of development" during which — as Shatz stretched her analogy still further — the child's brain "is almost running test patterns on all these connections, phoning home, essentially, to figure out which are the right phones to ring and which are the wrong and the incorrect connections are eliminated and the correct ones are actually strengthened and grow like mad." Also speaking at the conference was developmental psychologist Deborah A. Phillips, who related this process of wiring a child's brain to specific educational outcomes, stating that "neuroscience tells us" that young children in "sub-optimal child care environments" experienced adverse effects in "virtually every domain of development that we know how to measure, whether it's problem solving skills or social interactions or attention span or verbal development."[3] Hillary Clinton added, "Fifteen years ago, we thought that a baby's brain structure was virtually complete at birth. Now, we understand that it is a work in progress, and that everything we do with a child has some kind of potential physical influence on that rapidly-forming brain."[4] And President Clinton brought it all together when he announced plans to extend health care coverage and expand early educational opportunities for poor children. The key mission of the conference was unequivocal: use the new science of early childhood development to shine a spotlight on the deleterious impact that impoverished environmental conditions can have on a child's brain, especially during the first three years of life.[5]

However, there turned out to be a set of critical dilemmas associated with efforts to marshal neuroscientific and psychological data to advance liberal policies on early childhood intervention. First of all, as developmental psychologist Jerome Kagan noted, it was preposterous to presume that a "kiss, hug, lullaby, or scolding alters the child's brain in ways that will influence his future." Kagan advised that "every smile at an infant is not to be viewed as a bank deposit accumulating psychic dividends."[6] Philosopher of science John T. Bruer went further in his critique, stating that "neuroscience has discovered a great deal about neurons and synapses, but not nearly enough to guide educational practice." Bruer belittled the notion that the brain of a poor child suffered specifically due to social circumstances, writing that "normal children in almost any environment" acquire "sensory, motor, and working memory functions," no matter whether they are "children in affluent suburbs, children in destitute inner cities, [or] children in rural-pastoral settings throughout the world."[7] It was "selective, oversimplified, and interpreted incorrectly," Bruer wrote, to argue how "certain childhood experiences are necessary for optimal brain development."[8] He counseled that "we should be careful not to use neuroscience to provide biological pseudo-arguments in favor of our culture and our political values and prejudices." He also asserted that even the brain of a child in the worst conditions still "retains sufficient plasticity to compensate for deprivation" and will therefore regain "normal function if appropriate sensory experience occurs."[9] And in 2002, Harvard cognitive psychologist Steven Pinker cited Hillary Clinton's policy pronouncements based on childhood brain development as particularly boneheaded. Pinker wrote that there existed "no evidence that providing *extra* stimulation" enhanced the growth of a child's brain; he observed that "no psychologist has ever documented a critical period for cognitive or language development that ends at three." Pinker mockingly suggested that a reliance on sketchy brain science was unlikely to lead to effective public policy; it was simply not plausible that "the ills of the inner city" could be entirely "blamed on children's having to stare at empty walls."[10]

There were further dilemmas. *The Bell Curve*'s successful incitement of a fierce debate over race and intelligence was itself shifting the national conversation surrounding the political implications of human biology. The book had once more put genetics back on the table as an explanation for why there remained "cognitive differences between races." Moreover, it had advanced the idea that there existed a causal relationship between intelligence and class structure in the United States, with a "cognitive elite" rising steadily into the upper socioeconomic stratum, while a cognitive "underclass" slid ever further into the lower depths. It argued that "children unlucky enough to be born to and reared by unmarried mothers who are below average in intelligence" tended "to have low cognitive ability themselves" and so suffered "disproportionately from behavioral problems" and thus ended up "disproportionately represented in prison."[11] Herrnstein and Murray's logic may have infuriated critics because it blamed the poor for their own predicaments. But the critics often bought into its premise that it was necessary to use good science to fight bad science. Not unlike the liberal reaction to the "scientific" racist backlash against *Brown* four decades earlier — which had sought empirical data to buttress its case for justice in the form of raised IQs thanks to desegregation or enrichment programs — quite a few developmental psychologists and cognitive neuroscientists in the 1990s and early twenty-first century came essentially to argue that genuine neuroscientific evidence was the key to proving that *The Bell Curve* was pseudoscience.

In the quarter century since *The Bell Curve* first appeared, researchers have worked to identify a causal — and not merely correlational — relationship between socioeconomic status (SES) and cognitive function in the developing brains of young children. Much of this research has relied on its own version of a theory of neuroplasticity and has undertaken to prove that the plasticity of a developing brain renders a poor child *both* especially exposed to environmental harms *and* remarkably available to early educational enrichments. A major review in 2002 cited a wealth of data that SES was "associated with a wide array of health, cognitive and socioeconomic outcomes in children, with effects beginning prior to birth and continuing into adulthood."[12] Yet the review conceded that hard evidence still remained out of reach. There may be general acceptance that "growing up in poverty is associated with reduced cognitive achievement as measured by standardized intelligence tests," the journal *Brain Research* wrote in 2006, but "little is known about the underlying neurocognitive systems responsible for this effect." Still, the journal averred that "knowledge of the neurocognitive profile of poverty may have practical benefits" even before "the causal factors have been elucidated." Such a profile of poverty may provide "more specific targets for intervention programs, allowing us to more precisely address the neurocognitive vulnerabilities of at-risk children." Such a profile might "renew and expand our sense of societal obligation to poor children by reframing the problem as more than mere educational and economic opportunity, extending to the physical integrity of children."[13] But the problem had, in essence, already been reframed. A scramble to locate scientific proof became primary, even while researchers continued to argue for the *ethical* imperative to endorse early education intervention — even (or even especially) while causal factors remained unresolved.

The racial logic promoted by Herrnstein and Murray *was* insidious, and it needed to be opposed. But the noble enterprise of seeking to wipe away forever the fake science of *The Bell Curve* with the real science of early childhood development was to remain — and still remains — unrealized. Zombielike, *The Bell Curve*'s analysis refused to die, even as cognitive and developmental psychologists and their colleagues in the neurosciences have continued to pursue their goal with technologies and methodologies perhaps ill-suited to the task. In the meantime, more questions have arisen: If poverty does harm kids' neurocognitive abilities, and it is almost common sense to assume that it does, is it possible always to reverse the damages done or — conversely — to establish at what stage in child development it becomes too late to make a meaningful difference? Is neuroplasticity in essence a liberal's theory of brain development? Or: Could it be successfully refurbished to serve policies from the right? And: Would locating a causal connection between socioeconomic status and early cognitive development finally banish the beast of scientific racism once and for all — or might it only result in further unintended consequences?

Racial Differences in Intelligence (Redux)

Herrnstein and Murray's *The Bell Curve* appeared in the early fall of 1994 and swiftly rose to best-seller status, selling 400,000 copies in its first two months after publication. It received coverage in nearly every major newsmagazine and newspaper in the country; it was discussed on National Public Radio as well as on popular television news programs like *Good Morning America* and *Meet the Press*. The book also prompted an often passionate outpouring of counterarguments, especially for its contentions concerning "racial differences in intelligence."[14] For instance, in October 1994, *New York Times* columnist Bob Herbert wrote that *The Bell Curve* was "a scabrous piece of racial pornography masquerading as serious scholarship." He added that the book "strongly suggests" how "the disparity" in intelligence between whites and African Americans "is inherent, genetic, and there is little to be done about it." Herbert bitterly continued with what must have been one of the longest sentences ever to appear in the paper of record:

> I would argue that a group that was enslaved until little more than a century ago; that has long been subjected to the most brutal, often murderous, oppression; that has been deprived of competent, sympathetic political representation; that has most often had to live in the hideous physical conditions that are the hallmark of abject poverty; that has tried its best to survive with little or no prenatal care, and with inadequate health care and nutrition; that has been segregated and ghettoized in communities that were then redlined by banks and insurance companies and otherwise shunned by business and industry; that has been systematically frozen out of the job market; that has in large measure been deliberately deprived of a reasonably decent education; that has been forced to cope with the humiliation of being treated always as inferior, even by imbeciles — I would argue that these are factors that just might contribute to a certain amount of social pathology and to a slippage in intelligence test scores.[15]

Herbert's scathing takedown was to be only one of many that sought to eviscerate *The Bell Curve*. Many further critics emphasized especially what they saw as *The Bell Curve*'s extensive reliance on fraudulent evidence concerning a genetic basis for human intelligence.[16]

Not all commentators, however, were quite so dismayed or disgusted by *The Bell Curve*'s analysis. Some were more nuanced and tempered in their responses. Harvard sociologist Christopher Winship wrote (also in the *New York Times*) that *The Bell Curve* might have "serious flaws" but that it offered a number of "potentially valuable insights," including the book's assertion that "cognitive ability is largely immutable." Winship added that the book might be "overly pessimistic" but that "surely we would be naïve to think that simply increasing federal funding for early-childhood education" would "be sufficient to compensate for the increasing gap between the highly educated and the barely literate in U.S. society."[17] When the well-respected University of Chicago economist (and future Nobel Prize winner) James J.

Heckman reviewed the book, he observed that Herrnstein and Murray's core argument "fails" not least of all because its "central premise" was "the empirically incorrect claim that a single factor — *g* or IQ — that explains linear correlations among test scores is primarily responsible for differences in individual performance in society at large." But at the same time, Heckman acknowledged that *The Bell Curve*'s "challenge to contemporary assumptions about the malleability of human beings and the relative importance of environmental factors is courageous and long overdue."[18] Indicative and quite significant as well was the official report on *The Bell Curve* from the American Psychological Association. Drafted and unanimously approved by a committee of eleven prominent psychologists, the APA report sought neutrally to summarize every side of the issue of race and intelligence that *The Bell Curve* had thrust so dramatically into popular discussion. It critically noted that "all genetic effects on the development of observable traits are potentially modifiable by environmental input" and that it was "a common error" to conclude "that because something is heritable it is necessarily unchangeable." But the APA report discounted the view that a "differential between the mean intelligence test scores of Blacks and Whites" was merely a reflection of "differences in socioeconomic status." The report observed that it was possible that "environmental difference has caused the IQ difference" insofar as "growing up in the middle class produces higher psychometric intelligence than growing up poor." But the report stated that it was just as possible that "individuals can come to be in one environment or another because of differences in their own abilities." The APA report drily concluded, "Explanations based on factors of caste and culture may be appropriate, but so far have little direct empirical support."[19] Such assessments no doubt provided *The Bell Curve* a legitimacy both as science and as policy it would otherwise have been less likely to obtain.

Defenders of *The Bell Curve* seized the moment. On December 13, 1994, the *Wall Street Journal* published a full-page opinion piece pointedly titled "Mainstream Science on Intelligence." Signed by more than fifty "experts in intelligence and allied fields" and written by Linda S. Gottfredson, an educational psychologist at the University of Delaware, the statement offered stirring support for *The Bell Curve*'s several guiding principles, including that intelligence tests provided an excellent means of cognitive assessment; that IQ tests were not culturally biased against minorities; that differences in IQ did exist *within* racial or ethnic groups, but also that differences in IQ existed *between* racial and ethnic groups; that these group differences in intelligence were likely due far more to "heritability" and "genetics" and much less to "environment"; and that "the bell curve for whites is centered roughly around IQ 100," while "the bell curve for American blacks [is centered] roughly around 85." The *Wall Street Journal* statement asserted that there was a "practical importance" to the science of racial differences in IQ. A high IQ could not guarantee success in life, and a low IQ did not preordain defeat. But IQ scores were "strongly related" to "many educational, occupational, economic, and social outcomes." Employment that required "routine decision making or simple problem solving (unskilled work)" could be handled by individuals with lower IQs. But given that the modern world had grown "highly complex," this "complexity" meant that "a high IQ is generally necessary to perform well" in "the professions" and in positions of "management."[20]

The words "complex" and "complexity" had been important in *The Bell Curve*'s lexicon. Herrnstein and Murray had argued that the ability of an individual to grasp "complexity is one of the things that cognitive ability is most directly good for" and that the inescapable "complexity" of contemporary life meant that what was easier for "the cognitive elite" was becoming (more and) "more difficult for everyone else." The "everyone else," Herrnstein and Murray wrote, likely included "a large majority of the next generation of blacks in the inner city" who were "growing up without fathers and with limited cognitive ability." An increased segregation of persons not by race but by cognitive skills was inevitable, they warned, and they went on to announce that when "the ablest blacks" chose to flee "the inner city," it left "the inner city without its former leaders and role models." Increased isolation and dependency in these most impoverished — and least mentally competent — communities would ensue, while "the cognitive elite" (both white and minority) found itself in the role of having to establish "an expanded welfare state for the underclass that also keeps it out from underfoot."[21] This was, to be sure, a vividly dystopian — and racialist — vision of a near-future in the making.

Psychologist Gottfredson riffed freely on *The Bell Curve*'s logic in her own polemical writing. She grimly observed how "from birth to death, life continually requires us to master abstractions, solve problems, draw inferences, and make judgments on the basis of inadequate information." No surprise that "life is difficult at the low end of the IQ bell curve." Down at that "low end" were the residents in a "low-IQ White Appalachian community" *and* "Black inner-city special education students," all of whom had to struggle to complete their daily tasks. The aim of social policy had therefore to be a "narrowing" of what such individuals could be expected to learn and to accomplish in order to match their cognitive limitations. As for the children of the cognitive underclass, schools had to begin to use materials that were "the most essential" and the most basic — and presented "in the most accessible way"— so that these kids would be able to learn anything at all. Schools had drastically to terminate their tendency to "overpromise" an ability to educate those who were not really educable. Gottfredson concluded with the harsh punch line, "Civil rights advocates resolutely ignore the possibility that a distressingly high proportion of poor Black youth may be more disadvantaged today by low IQ than by racial discrimination, and thus that they will realize few if any benefits (unlike their more able brethren) from ever-more affirmative action."[22] And so at the edges of her arguments — or perhaps at their core, it was not consistently clear — always lurked the not-so-hidden subtext of racial differences in intelligence. The second constant was the message that the government should stop directing money to the poor and retract the welfare state (while associating that welfare state with — implicitly undeserving — blackness).

Gottfredson acknowledged that "intelligence" in humans was likely "multiple." But she always returned — as had Herrnstein and Murray — to the singular significance of general intelligence (or "g"). It was not in the end a terribly different argument from the one that educational psychologist Arthur Jensen had been making (and making repeatedly) since the late 1960s, and it ultimately sought to diminish the role that more squishy and "less cognitive traits such as personality or interests" played in terms of predicting whether an individual would likely make it in work and in life. As Gottfredson bluntly put it in 1997, "There are many other valued human traits besides g, but none seems to affect individuals' life chances so systematically and so powerfully in modern life as does g." Gottfredson came to summarize the human condition like this (with italics): "*Higher Levels of* g *Are Required up the Occupational Ladder.*"[23]

Much of the disruptive appeal of *The Bell Curve* for its defenders resided in the fact that it so self-reflexively challenged post-1960s racial liberalism. The book's advocates cheered how Herrnstein and Murray disobeyed cultural conventions and used (what they said were) empirical verities to debunk liberal orthodoxies. What *The Bell Curve* had achieved — or so its supporters announced — was that it took "mainstream science" to demonstrate (what it said was) a simple fact: All races were *not* created cognitively equal. The book's defenders liked to argue that the book was breaking taboos that needed to be broken — even though these were quite obviously "taboos" that had *always* been routinely broken in the length and breadth of U.S. history. (Before the 1960s, when had there been an age when it actually was *not* possible for social scientists to argue openly "from within a biological framework" for "the innate inferiority of people of African descent"?)[24] The terrible and tragic paradox of *The Bell Curve*'s lasting impact on policy discussions of race and intelligence was that it served to promote age-old racial prejudices while insisting that it was just telling it like it really was.

Again: Gottfredson may have put it best. She argued that the scientific evidence presented in *The Bell Curve was* real — and *that* had been its problem. What had caused all the fuss was that Herrnstein and Murray refused to endorse (what Gottfredson had named even before *The Bell Curve* appeared as) the "falsehood, or 'egalitarian fiction,'" that "racial-ethnic groups never differ in average developed intelligence (or, in technical terms, *g*, the general-mental ability factor)." Unless a psychologist or anyone else dutifully genuflected before this "falsehood," he or she confronted censorship or sanctions both "covert and overt." There were strict limits on what one could say in public forums in the United States of the 1990s — especially when what was being said involved "certain truths about racial matters today." These tough restrictions on free expression pressured social scientists into "subordinating scientific norms to political preferences," a twisted state of affairs that had resulted in "our current pseudo-reality on race."[25] Gottfredson also pithily wrote, "Nothing frightens the liberal mind more than the prospect of inherited differences in intelligence between the races."[26] This smirking aphorism neatly summarized the plight of *anti*-racist scientists and liberal policy makers by the end of the century, as right-wing activists reveled in the ideological conflagration *The Bell Curve* had sparked so furiously to life.

Brains Built over Time

The Bell Curve — and everything it rapidly came to represent — provides the essential context for comprehending the ensuing intensity of efforts to articulate an alternative perspective that underscored the plasticity of brain development in early childhood. These efforts to formulate an alternate perspective were what had animated Hillary Clinton's statements in *It Takes a Village* in 1996. During and after the Clinton White House conference in 1997, these efforts also became the subject of a far larger and more sustained media blitz aimed at educating the general public specifically about the political implications of the concept of neuroplasticity in young children. A nationally broadcast ABC-TV program, *I Am Your Child*, aired that spring, coinciding with the release of "Your Child: From Birth to Three," a special edition of *Newsweek* magazine (which would sell more than a million copies). The special edition of *Newsweek* stated how pediatricians could now watch an infant's brain grow — courtesy of positron-emission tomography — which made it possible to witness in real time as "the regions of a baby's brain turn on, one after another, like city neighborhoods having their electricity restored after a blackout."[27] *Rethinking the Brain*, a companion text to the Clinton conference, discussed the brain's remarkable capacity to adapt, even as it counseled that there were "sensitive periods during which the brain is particularly efficient at *specific* types of learning," and so therefore "the brain's plasticity presents us with opportunities and responsibilities."[28] In its discussion of brain plasticity, *Time* magazine too laid bare its progressive leaning when it stressed how there existed an "urgent need, say child-development experts, for preschool programs designed to boost the brain power of youngsters born into impoverished rural and inner-city households." *Time* advised, "By the time of three, a child who is neglected or abused bears marks that, if not indelible, are exceedingly difficult to erase."[29] Collectively these news reports deliberately sought to turn political discussions away from the reasoning in *The Bell Curve* that used biology to reduce educational assistance for disenfranchised children and toward a liberal logic that invoked biology to increase funding for these same children. Biology played a central role in the political calculus on *both* sides of the equation.

Popular accounts that pushed a theory of neuroplasticity — and the ways in which brain plasticity helped us grasp the dire consequences of childhood poverty — employed a range of vivid metaphors, even as their message remained much the same. There was the brain-as-computer metaphor, as when science journalist Sharon Begley observed, "It is the experiences of childhood, determining which neurons are used, that wire the circuits of the brain as surely as a programmer at a keyboard reconfigures the circuits in a computer." Begley got quickly back on point when she added, "Which keys are typed, which experiences a child has[,] determines whether the child grows up to be intelligent or dull, fearful or self-assured, articulate or tongue-tied."[30] There was the brain-as-orchestra metaphor: "A baby is born with more than 100 billion brain cells," wrote *Education Week*, some of which were "hardwired" at birth. "But the rest are just waiting to be hooked up and played like orchestra instruments in a complex musical composition."[31] And there was the most important "windows of opportunity" metaphor. "Windows of opportunity" came to dominate discussion about early child development by the middle years in this "Decade of the Brain," with talk of how "different circuits" in a child's brain "are most sensitive to life's experiences at different ages," and therefore children had distinct "windows of opportunity" in terms of their capacity to learn.[32] "Windows of opportunity" opened in "the first three years," when "the foundations for thinking, language, vision, attitudes, aptitudes, and other characteristics are laid down," Pulitzer Prize–winning science journalist Ronald Kotulak wrote in his influential book *Inside the Brain*. "Then the windows close, and much of the fundamental architecture of the brain is completed."[33] And mixing metaphors a bit, a title in the journal *Young Children* effused in 1997, "New Brain Development Research — a Wonderful Window of Opportunity to Build Public Support for Early Childhood Education!"[34] In all of these rhetorical framings, the brain was always also plastic, but not for long (and never forever). To cite a final metaphor: a child's brain was "a super-sponge," but it was a "super-sponge" that soon lost its absorbency.[35]

That the child's brain was plastic and that it developed in relation to environmental factors were hardly novel perspectives in the 1990s. Neuropsychologist Donald O. Hebb had argued already in the late 1940s that "neural change" could be "induced by experience"; Hebb had posited that experience could modify the structure of the brain not only in animals but also in humans.[36] By the early 1960s, Berkeley neuroanatomist Marian C. Diamond along with her colleagues David Krech and Mark R. Rosenzweig, biological psychologists, announced "that increasing the environmental complexity of rats resulted in measurable changes in brain chemistry and in brain weight."[37] By 1964 this same team of researchers, plus psychologist Edward L. Bennett, reported on what they now unequivocally labeled the "anatomical plasticity" of the brain.[38] It appeared that an animal's brain was "most plastic" while young and increasingly less so as it matured into adulthood.[39] By the 1970s, and building directly on the work of Diamond, Krech, and Rosenzweig, developmental psychologist William T. Greenough and his colleagues found evidence of significantly greater synaptic growth in rats weaned in "complex" rather than isolated environmental conditions.[40] By the 1980s, Greenough extrapolated his findings to humans, postulating that the development of new synapses in young children was either *"experience expectant"* (that is, "designed to utilize the sort of environmental information that is ubiquitous" to all) or *"experience dependent"* (that is, unique "in both timing and character among individuals"). The synapses of a child's brain were initially overproduced, only subsequently to be pruned down in relation to specific experiences that determined "the pattern of connections that remain." In a widely cited essay from 1987 in the journal *Child Development*, Greenough and his colleagues highlighted the "active formation of new synaptic connections in response to the events providing the information to be stored," as well as the idea of "sensitive periods" when such processes were most likely to occur.[41] Greenough's model that spoke of "sensitive periods" in brain plasticity became a crucial component of arguments for the "neurobiological bases of early intervention."[42]

With increasing force, psychologists and other commentators began to foreground what they stated were the profound implications of research on brain plasticity for the plight of children from disadvantaged circumstances. As discussed in chapter 1, psychological evidence that "pre-school enrichment" improved scholastic outcomes for impoverished youth had fueled initial support in the mid-1960s for early education programs like Head Start. By the mid-1990s, as the fiscal tug-of-war with skeptical conservatives continued, there was said to be *neuroscientific* evidence to support the greater investment of federal funding in these intervention programs. In 1996 developmental psychologist Craig Ramey—whose research would be prominently cited by Hillary Clinton that same year—argued that policy makers would never have been able to ignore the level of knowledge about the vulnerability of poor children's neural development "if we had a comparable level of knowledge with respect to a particular form of cancer, or hypertension, or some other illness that affected adults," but since "young children don't vote and they can be sort of kept hidden for a while until they show up in the school system failing miserably," their difficulties could be more easily neglected.[43] Ramey, along with his wife and frequent collaborator, developmental psychologist Sharon Landesman Ramey, did not hesitate to single out Herrnstein and Murray for critique. Among other things, they wrote that emerging "empirical evidence on biobehavioral effects of early experience" should finally lay to rest the "erroneous explanation" for "educational and cognitive inequalities" that remained "alive and socially influential even today" as demonstrated by "the popularity" of *The Bell Curve*.[44] A *New York Times* editorial in 1997 made its allusion to Herrnstein and Murray only slightly more oblique when it noted how "many people have accepted the notion that the brain has been genetically determined by the time a baby is born" but that "research by neuroscientists, however, shows that after birth, experience counts even more than genetics."[45] *The Bell Curve*, or at least what *The Bell Curve* was said by its critics to represent, lingered in late-1990s discussions on how (and why) neuroplasticity should inform policies on early intervention.

In hindsight, it is difficult not to read the fin de siècle move that made neuroplasticity a centerpiece in arguments favoring early intervention in historic relationship to *The Bell Curve*. Herrnstein and Murray had said nothing whatsoever about synaptic development in early childhood. Plasticity was not a concept they considered relevant or worthy of attention; the pioneering research of Hebb, Diamond, Krech, Rosenzweig, Greenough, and anyone else who had explored the interplay of early experience and brain development all went unmentioned in *The Bell Curve*. In turn, however, to establish that the science of neuroplasticity was crucial for an understanding of early childhood development was manifestly intended to demonstrate by default that *The Bell Curve*'s analysis, its archive, and everything it stood for had been both selective in its choice of evidence and slanted by its ideological intent.

Meanwhile, a theory of neuroplasticity began to have direct effects on early education policies. Governors in at least a dozen states requested that more monies be allocated for early education programs, and California governor Pete Wilson cited specifically the science of early brain development when he asked that more than $740 million be added to the state budget for child assistance.[46] Also the Early Childhood Development Act of 1997, whose key sponsor was Senator John Kerry of Massachusetts, stated that "new scientific research shows that the electrical activity of brain cells actually changes the physical structure of the brain itself and that without a stimulating environment, a baby's brain will suffer." This federal legislation appropriated $1.5 billion each fiscal year to support "good childhood learning services," especially for the 25 percent of children in the United States who were living in poverty.[47] Taken together, these initiatives came to represent — as worthy as their ends might otherwise have been — a full-scale "politicization of research on early brain development and child development."[48]

A politicization of early development research proceeded apace into the first decade of the new century. In 2000, the National Research Council and Institute of Medicine jointly published *From Neurons to Neighborhoods: The Science of Early Childhood Development*. *From Neurons to Neighborhoods* had been first proposed in early 1997 by the National Academy of Sciences' Board on Children, Youth, and Families. At that time, the board had established the Committee on Integrating the Science of Early Childhood Development, charging it with the task of preparing a report that would "review what is known about the nature of early development and the influence of early experiences on children's health and well-being, to disentangle established knowledge from erroneous popular beliefs or misunderstandings, and to examine the implications of the science base for policy, practice, professional development, and research."[49] A member of the board of directors for the child advocacy organization Zero to Three, pediatrician Jack Shonkoff, chaired the committee and coedited the report with developmental psychologist Deborah Phillips, who had spoken at the Clinton White House conference. Other committee members included psychologists Alicia F. Lieberman and Megan Gunnar — both of whom served on Zero to Three's board of directors — as well as William Greenough and several others with expertise in the science of early childhood development. The committee received funding from numerous agencies and programs dedicated to education and public health. These included the National Institute of Mental Health, the Centers for Disease Control and Prevention, the U.S. Department of Health and Human Services, and the U.S. Department of Education.

From Neurons to Neighborhoods was the most comprehensive study of its kind. In more than 400 pages of text, and an additional 150 pages of references and appendices, it sought to make sense of new directions in neuroscience and to integrate this often complex research into an accessible presentation of how to formulate more successful early-childhood-related policies and services, especially for underprivileged families and children. The report argued that "early experiences clearly affect the development of the brain" and that "poor nutrition, specific infections, environmental toxins, and drug exposures, beginning early in the prenatal period, as well as chronic stress stemming from abuse or neglect throughout the early childhood years and beyond," all posed threats "to the developing central nervous system." Throughout, the text engaged a dynamic theory of child development by emphasizing "the interplay of nature and nurture." The report argued that it was scientifically fallacious to embrace "the misleading and tired old nature-nurture debate." Instead it was critical to reconceive child development within a "transactional-ecological" framework. This meant taking seriously the idea that each child was "unique" and each environment was experienced by a child "in unique ways." The report's authors put it like this: "In simple terms, children affect their environments at the same time that their environments are affecting them." The report stated, "It is not nature

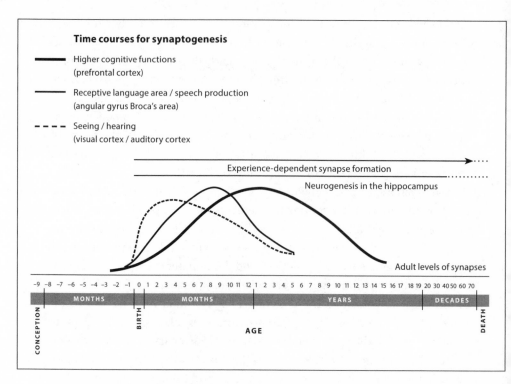

FIGURE 14. This graph by pediatrician Charles A. Nelson appeared in Jack P. Shonkoff and Deborah A. Phillips's *From Neurons to Neighborhoods*. It served to illustrate how "the disproportionate attention being given to the period from birth to 3 years begins too late and ends too soon." Reprinted by permission of Charles A. Nelson III.

versus nurture, it is rather nature *through* nurture." And it added, "Nature is inseparable from nurture, and the two should be understood in tandem." The report's pithy title was elucidated in this context: "At every level of analysis, from neurons to neighborhoods, genetic and environmental effects operate in both directions."[50]

From Neurons to Neighborhoods did remain cautious in its pronouncements. It expressed concerns that "to disentangle established knowledge from erroneous popular beliefs or misunderstandings" was not an uncomplicated task. It strove at all points to be clear and unsensational; it granted the inadequacy of "the recent focus on 'zero to three' as a critical or particularly sensitive period" (see figure 14). Also significant was that the report conceded that the efforts of early educational intervention could never be expected—

in some mechanistic way—to raise IQs or secure positive lifelong outcomes for disadvantaged children. The report admitted that a great deal remained unknown. But it stated that "one of the most consistent associations in developmental science is between economic hardship and compromised child development." And it declared as well that it remained the country's "ethical and moral" responsibility to assist its most vulnerable children. "How can the nation use knowledge to nurture, protect, and ensure the health and well-being of all young children as an important objective in its own right, regardless of whether measurable returns can be documented in the future?" the report inquired, however rhetorically. At the same time it openly criticized *both* sides of the debate over early childhood intervention for perpetuating what it said was a "siege mentality" between "those who support massive public investments in early childhood services and those who question their cost." It declared that "both perspectives have merit," even while it underscored (again) that there was "massive scientific evidence that early childhood development is influenced by the environments in which children live." The report concluded that science alone would not settle this "conflict between advocates and skeptics." The science of early childhood development was simply not irrefutable—at least not yet. Instead the report affirmed that early intervention could be "helpful in shifting the odds toward more optimal pathways of later growth." The fact of neuroplasticity in human development was cause for optimism—even while it also generated cause for apprehension. "Plasticity is a double-edged sword that leads to both adaptation and vulnerability," the report observed, adding that "plasticity cuts both ways" due to the myriad of "environmental conditions" that could leave young children "simultaneously vulnerable to harm and receptive to positive influences."[51]

From Neurons to Neighborhoods did not mention Herrnstein or Murray by name. But the report's "core story of child development" did emphasize the malleability of human development in response to particular life experiences.[52] It did speak early and often of how "processes of brain development that were traditionally regarded as genetically hard-wired" should be understood as wholly dependent "on an exquisitely coordinated dance between experiential catalysts and the hereditary design for brain growth." It observed repeatedly that "the mechanisms of neurodevelopment are designed specifically to recruit and incorporate a broad spectrum of experience into the developing architecture of the brain," adding that, in fact, "the brain's ongoing plasticity enables it to continually resculpt and reshape itself in response to new environmental demands well into adulthood."[53] These perspectives, taken together, served as counterpoints to *The Bell Curve*'s recurrent message that there was an unmistakable—and potentially irremediable—differential between whites and African Americans in standardized IQ test scores.

The class-based themes and color-blind arguments in *From Neurons to Neighborhoods* came to dominate the new century's literature on the science of early childhood. The book would be cited thousands of times in the decades to follow. Harvard University established a National Scientific Council on the Developing Child in 2003 chaired by Jack Shonkoff and with a list of members that drew extensively from the committee that had prepared *From Neurons to Neighborhoods*. The council effectively picked up right where the 2000 report left off. From the start, the council made evident its commitments via a series of working papers with titles like "Excessive Stress Disrupts the Architecture of the Developing Brain" (2005), "The Timing and Quality of Early Experiences Combine to Shape Brain Architecture" (2007), and "Early Experiences Can Alter Gene Expression and Affect Long-Term Development" (2010). "Brains are built over time" became the mantra of the movement that took the science of early child development to be a key resource in the policy struggles over offering more assistance to young children living in poverty.[54] In discussing the consequences of "toxic stress" on young children, Shonkoff told the *New York Times* in 2013, "What the science is telling us now is how experience gets into the brain as it's developing its basic architecture and how it gets into the cardiovascular system and the immune system."[55] And in 2015, a *New Yorker* article, "What Poverty Does to the Young Brain," underscored the point that poverty itself was a "sort of neurotoxin." Whereas children from more affluent families developed brains with a greater surface area "and a more voluminous hippocampus," poorer children had less brain surface—and the poorer they were, the more dramatic the differences turned out to be. Studies demonstrated that poorer children were more likely to experience cognitive problems into adulthood. "Over the past decade," the *New Yorker* article stated, "the scientific consensus has become clear: poverty perpetuates poverty, generation after generation, by acting on the brain." Or as the article also opined, "In other words, wealth can't necessarily buy a better brain, but deprivation can result in a weakened one."[56] These were the arguments Hillary Clinton had presented at the White House conference on Early Childhood Development and Learning in 1997. They remained the arguments being made close to twenty years later.[57]

Opponents of *The Bell Curve* came in these years to see brain science as having made their case for them. They began to sound as if the war had been won. In 2007, when the journal *Educational Leadership* noted the increasingly "robust" findings that demonstrated "that, under the right conditions, early intervention can dramatically improve the odds for children at risk," it could not resist taking a jab at Herrnstein and Murray for their "profound skepticism that any form of education intervention can alter the cumulative negative toll that poverty and other disadvantages take on the development of young children." The journal went on assuredly to say that thanks to the new science of child development, it appeared possible once and for all "to explode the myth that nothing works."[58] But was *The Bell Curve* truly dead?

The Neglect of the Gifted

A problem for critics was that *The Bell Curve* had not quite said what they said it said. Herrnstein and Murray had among other things argued that any law that clustered individuals together, racially or otherwise—such as "anti-discrimination legislation" (for example, affirmative action in the workplace or education)—represented a "leaking poison into the American soul." They had stressed that a society that did not acknowledge the key role played by cognitive ability was a society that promoted all manner of inadvertent inequities. They did state that it would be foolish to believe that "government activism or other strategies based outside the home" could ever really help children from "the underclass"—since their mothers were often "incompetent" and "incompetent mothers are highly concentrated among the least intelligent, and their numbers are growing."[59] And they had condemned the federal government for squandering vast sums of money year after year in its fruitless pursuit to help the children of the poor.[60]

But *The Bell Curve* had not discounted the impact of environmental factors on intelligence, as its critics often assumed. Herrnstein and Murray liked very much the story of "two handfuls of genetically identical seed corn," where one handful is planted "in Iowa, the other in the Mojave Desert," only then to "let nature (i.e. the environment) take its course." Herrnstein and Murray continued: "The seeds will grow in Iowa, not in the Mojave, and the result will have nothing to do with genetic differences." They pointed out, "The environment for American blacks has been closer to the Mojave and the environment for American whites has been closer to Iowa." (It was a peculiar and terrible irony that also white liberals had not infrequently in the postwar

decades used environment-focused arguments with pathologizing implications.) Moreover, Herrnstein and Murray repeatedly stated that "environment" played a part in shaping racial differences in intelligence, even as they
insisted that "genes" played a role as well. They wrote, "It seems highly likely
to us that both genes and the environment have something to do with racial
differences. What might the mix be? We are resolutely agnostic on that issue;
as far as we can determine, the evidence does not yet justify an estimate."[61]
(It was certainly a claim too vague to be completely contradicted.)[62]

In truth, environment mattered greatly to Herrnstein and Murray. This
was evident not just in their vignettes about poor children who suffered
"inadequate nutrition, physical abuse, emotional neglect, lack of intellectual stimulation, a chaotic home environment — all the things that worry
us when we think about the welfare of children."[63] Environment factored
heavily as well in their discussion of children from the "cognitive elite." For
despite a serious fondness for the genetics of human intelligence, Herrnstein
and Murray nonetheless strategically observed that these children had to
see their talents actively nourished. Cognitive brilliance might be innate,
but — like seed corn — it required the educational equivalent of Iowan soil.

In short, it was no contradiction that *The Bell Curve* promoted policies that
called for more funding to be funneled to America's high-IQ (and, incidentally or not, often white) students.[64] *The Bell Curve* condemned "the neglect
of the gifted." It cited federal budget statistics that saw more than 90 percent
of educational funding going to "the disadvantaged," while a mere tenth of 1
percent went to "programs for the gifted." It expressed horror at these numbers, and it urged policy makers to radically reallocate federal funds to "intellectually gifted" children who, Herrnstein and Murray argued, should be
permitted to reap the rewards of their own intelligence —"not because they
are more virtuous or deserving but because our society's future depends on
them."[65] Until such time as these monies were spread more evenly across the
children of the bell curve, there would continue to be a public school system
in the United States that failed children on nearly every front — neither managing to assist the disadvantaged child nor serving properly to encourage the
brightest children so that they might develop their fullest potential.

From this perspective, the concept of brain plasticity in early childhood —
rather than being seen as an obstacle for the political right to evade — came
over time actively to serve a right-wing agenda. This did not happen over-
night. As mentioned, *The Bell Curve* had sidestepped brain plasticity as a
topic. So too had a major two-volume policy document, *A Nation Deceived:*

The Schools Hold Back America's Brightest Students (2004), also known as "The Templeton National Report on Acceleration" (since it had been produced with support from the John Templeton Foundation). *A Nation Deceived* put its argument succinctly: "In every state, in every school, in huge cities, and in tiny farm communities, students are ready for much more challenge than the system provides." It provocatively continued, bringing not only race but also the only recently attended-to group of the disabled into the discussion: "While disabled students' rights to an appropriate education are protected by laws, there is little legal protection for the gifted. In most states, there are no laws mandating appropriate educational interventions for children who sit, underchallenged, in classrooms year after year." It went on indicatively—and remarkably—to declare "that 50 years after *Brown vs. Board of Education*, our country still has not achieved equality in the classroom. Brown [*sic*] began the journey to legally end grouping by skin color. Today, altering attitudes about acceleration is a journey to end grouping by birth date."[66] Still, there was no mention of neuroplasticity in *A Nation Deceived*.

Right-leaning commentators warmed only slowly to a perspective that brain plasticity might be friend and not foe. And this had everything to do with an intention to promote policies (and enhance funding) for gifted children. In 2006, when researchers observed that it was time for "a paradigm shift" in their field of gifted education, they noted that it was no longer possible to think of "some children as innately having brains that work unusually well." It was far more appropriate to conceptualize "intelligence as a developmental phenomenon that is importantly dependent on a child's opportunities to learn." They added, "The more we learn about brain development, especially neural plasticity, the clearer it is that we have to be careful when setting limits on people's potential for learning."[67] Likewise a 2008 anthology on gifted education outlined the "practical implications of the research on the neural bases of intelligence": "Providing an enriched environment is essential to the development of gifted and talented individuals, and it has profound consequences on brain maturation and subsequent neuroplasticity."[68] These points resembled those made in *From Neurons to Neighborhoods*—excepting the very different category of children it aimed to assist. Then Charles Murray in 2008 weighed in, writing that "important evidence has been found for the plasticity of certain mental processes, especially during infancy and early childhood." For Murray the science of brain plasticity needed no longer to be viewed as a threat to the Right's education policies; on the contrary, Murray invoked plasticity in the midst of a jeremiad against an "educational system that cannot make itself talk openly about the implications of diverse educational limits," an educational system that perpetuated "a fog" of "well-intended egalitarianism" and "educational romanticism." Murray once again — only now with plasticity on his side — closely hewed to *The Bell Curve*'s core analysis, insisting that education policy in America currently failed *all* students — not only the students who were "below average" but also "those who are lucky enough to be academically gifted" and who "will play a crucial role in America's future."[69]

And so a focus on environment and brain plasticity came to be used to advance policies for gifted education — and to undermine the position that only disadvantaged children required enrichment. In 2011, when the journal *Psychological Science in the Public Interest* announced (in italics) that "*both cognitive and psychosocial variables are malleable and need to be deliberately cultivated*," it did so in the midst of a case that there existed tremendous "resistance to gifted education by policymakers" and that "gifted children, regardless of the conditions under which they go to school or the economic status of their families, are not an educational priority and are assumed to be sufficiently capable of learning under most circumstances."[70] Neuroplasticity may have once been a pet theoretical concept adduced by liberal policy makers and politicians in order to make a case that more monies be directed to impoverished children, but no longer. Now plasticity could be appropriated — and applied — as a rationale to advance policies *across* the political spectrum.

Poverty and the Developing Brain

As for poverty's effects on the developing brain, the results from brain science remained inconclusive. An introduction to a special issue of the journal *Developmental Science* in 2013 on the "neurocognitive consequences of socioeconomic disparities" could only speak of an "emerging consensus that brain regions with a longer developmental time course, including prefrontal regions, may be more susceptible to variation in environmental input." Yet the introduction still contended that it was "time now to begin to harness the promise of developmental cognitive neuroscience research and to use it to ameliorate the effects of poverty on child development."[71] In 2015 *Developmental Psychobiology* reported the results of a study that found "socioeconomic disparities in neurocognitive development," especially in language and memory. But it acknowledged, "Of course we cannot say with certainty whether the associations reported here are indicative of causal relations."[72] That same year *JAMA Pediatrics*, seeking to link "child poverty, brain development, and academic achievement," found evidence that *"suggests* that specific brain structures tied to processes critical for learning and educational functioning (e.g. sustained attention, planning, and cognitive flexibility) are vulnerable to the environmental circumstances of poverty."[73] Meanwhile, the hunt continued for firm evidence that early educational intervention improved cognitive function for children of the poor. Developmental psychologist and cognitive neuroscientist Kimberley G. Noble wrote in the *Washington Post* in 2015, "The political battles for major expansion of these types of programs are unlikely to be won until we can provide hard scientific proof of their effectiveness. Until then, we need to do all we can to support policies that offer our most vulnerable children the best chance of reaching their full potential."[74] And in 2016 the journal *Pediatrics* also described research on interventions that might undo the neurocognitive damages of poverty to the developing brain as still very much "in its infancy," adding, "Perhaps most urgently, experimental studies that assess the impact of changing SES on brain development are needed to determine causal links."[75] However inadvertently, these statements were conceding quite a lot about the fragility of their own case.

Worse, was the search for "hard scientific proof" an instance of needing to be careful of what you wish for? In 2003, journalist Ann Hulbert had already observed that when politicians like the Clintons emphasized how "young brains subjected to deprived conditions" ended up "pickled in stress hormones" and "stripped of synapses" and thus became "irrevocably damaged" unless they were provided early interventions to assist them, such argumentation could well serve to "inspire a liberal social agenda." However, Hulbert cautioned, it remained equally possible that a "reading of the data" could "all too easily fuel defeatism" and "just as readily be invoked in the service of a deeply pessimistic position that was not at all what they intended."[76] Hulbert's insight was to prove prescient. In 2009, when University of Pennsylvania cognitive psychologists Daniel A. Hackman and Martha J. Farah surveyed the available neuroscientific literature on the effects of poverty on the developing brain, they determined that "the environments and experiences of childhood in different socioeconomic strata are at least in part responsible for different neurocognitive outcomes for these children." They took it as an article of faith that childhood poverty decreased language function and had an injurious impact on executive function. They accepted that the brain of a poor child frequently differed from the brain of a child from higher socioeconomic ranks. They emphasized the vital need for further research on "the neurocognitive outcomes" for poor children because "the better our ability to test hypotheses giving rise" to "disparities in cognitive function" as a result of lower socioeconomic status, "the more rationally we can design programs for prevention and remediation." Yet at the same time, Hackman and Farah acknowledged a potential problem that confronted research into the ways that poverty in childhood could be used as a "predictor of neuro-cognitive performance." They wrote, "Although the cognitive neuroscience of SES has the potential to enable more appropriately targeted, and hence more effective, programs to protect and foster the neurocognitive develop-ment of low SES children, it can also be misused or misunderstood as a rationalization of the status quo or 'blaming the victim.'" They added, "This has precedent in social science research on SES, in which characteristic dif-ferences between individuals of higher and lower SES have been used by some to argue that low SES individuals are intrinsically less deserving or less valuable members of society. The biological nature of the differences documented by cognitive neuroscience can make these differences seem all the more 'essential' and immutable."[77]

Put another way, then, if the terms of debate became set as principally biological, would solid proof that poverty damaged the cognitive skills of children of low socioeconomic status necessarily result in policy argu-ments for early intervention? The specter that a "possibility of 'real' deficit" in neurocognitive development might afflict poor children, and that this deficit might result "in irreversible neurocognitive impairment," haunted the journal *Frontiers in Human Neuroscience* in 2012.[78] So also in *Nature Neuroscience* in 2015: it was a worrisome concern, as a team of more than two dozen researchers wrote, that while "socioeconomic disparities are as-sociated with differences in cognitive development," it needed to be noted, "as a final point," that "our results should in no way imply that a child's socioeconomic circumstances lead to an immutable trajectory of cognitive or brain development." They added this disclaimer: "As such, many lead-ing social scientists and neuroscientists believe that policies reducing family poverty may have meaningful effects on children's brain functioning and cognitive development."[79] This became the new refrain — made often only in passing — that while the goal remained to "prove" that low-SES children experienced cognitive damage, this evidence should not be distorted in order to argue for an "irreversibility that could stigmatize children in poverty."[80] Rather than calling for remediating childhood poverty as a baseline social and political goal, commentators often got tangled in their efforts to adduce biological evidence while worrying that it could be turned against their cause.

Conclusion

On February 12, 2013, at the start of his second term as president, Barack Obama announced in his State of the Union address a major initiative to fund early childhood education. These funds, Obama declared, would seek to guarantee public preschool programs for children in families at or below the poverty line. Obama declared to Congress, "Every dollar we invest in high-quality early childhood education can save more than seven dollars later on — by boosting graduation rates, reducing teen pregnancy, even reducing violent crime." And he added how "studies show students grow up more likely to read and do math at grade level, graduate high school, hold a job, [and] form more stable families of their own" in "states that make it a priority to educate our youngest children."[81] In the days that followed, there were more details concerning this new "preschool for all" initiative. It would extend federal funding to all fifty states and would target four-year-old children in families at or below 200 percent of the poverty line. It would seek to provide high-quality programs through the use of "instructors with the same level of education and training as K–12 instructors." And it would see "a massively expanded Early Head Start program" aimed in particular at those most "vulnerable children ages 0 to 3."[82]

The Obama administration's budget proposal for 2014 followed through on its initiative to expand funding for early childhood education. It announced the introduction of a competitive award program, Preschool Development Grants, to assist with the establishment of high-quality preschools. It increased funding to Early Head Start. It supported childcare subsidies for low-income families. In sum, it sought to allocate $75 billion over the course of the next decade "to help low income parents meet their children's early education needs."[83] And to bolster the scientific rationale for Obama's universal preschool initiative, the White House website elaborated on the president's remarks by citing the neuroscience of early brain development:

"Research has shown that the early years in a child's life — when the human brain is forming — represent a critically important window of opportunity to develop a child's full potential and shape key academic, social, and cognitive skills that determine a child's success in school and in life."[84]

However, a Republican-controlled Congress showed no interest in advancing Obama's universal preschool initiative. In 2013, not only was the initiative blocked, but there were additional sharp spending cuts made to Head Start. In 2014, the Democrats successfully pushed to increase funding to Head Start and Early Head Start, and some $250 million were allocated finally to commence the "pre-school for all" initiative; later that year, the Department of Education introduced the Preschool Development Grants program, designed to provide "high-quality early learning programs" for "all children, but especially children from low-income families."[85] But in 2015 Republicans moved to "zero out" funding for the Preschool Development Grants and cut federal support to "over 200 high-need communities across 18 states that span the geographic and political spectrum."[86] These bills further proposed significant reductions to Head Start services.

A week before Obama left office in January 2017, *Education Week* tersely summarized the likely legacy of Obama's K–12 agenda, and especially the fate of his administration's universal preschool initiative. The newspaper observed that Obama had "pitched a big investment in early-childhood education in his 2013 State of the Union address" but that a Republican-controlled House of Representatives had "never seriously considered it."[87] Only days before the Obama administration left office, the Department of Education and the Department of Health and Human Services issued a joint policy statement that included several references to decades of scientific research on early childhood. In this administration's swan song to the nation, these departments declared, "The first three years in a child's life is a critical period for brain development." The joint statement continued, "The brain is strengthened by positive early experiences," whereas "negative experiences and toxic stress in the early years can disrupt brain development and can have significant irreversible damage on the immature brain." Children in poverty were "most susceptible to suffering the effects of negative experiences," the statement declared, a fact that only underscored the need for a range of "effective early childhood interventions" to assist especially the most vulnerable children and families in order to "significantly improve school readiness, academic success, development, and overall well-being."[88] It remained wholly unclear, however, what role (if any) the accumulated evidence from brain science would play in shaping the educational policies of the next administration.[89]

Meanwhile, moreover, and in the midst of all these other trends and uncertainties, a further key development was becoming increasingly apparent: a growing number of low-income children of color in public schools were ever more routinely being directed into the juvenile justice system. Juvenile arrests of African American youth had climbed exponentially in the first decade of the twenty-first century, as more public school districts nationwide enforced increasingly punitive policies in an attempt to maintain discipline. These "zero-tolerance" measures disproportionately affected poor children of color. Ill-defined charges like "disorderly conduct" or "disturbing school" led to an expanding presence of police in schools as well as to arrests of African American and Latino students at rates that far outstripped those of white pupils—even when charged with the same infractions. The same was true for rates of school suspensions, with black youth suspended far more often than white students for equivalent disciplinary charges. The consequences of arrests or suspensions were often dramatic. "Exclusion from the classroom for even a few days disrupts a child's education and may escalate misbehavior by removing the child from a structured environment, which gives the child increased time and opportunity to get into trouble," wrote legal scholars Catherine Y. Kim, Daniel J. Losen, and Damon T. Hewitt in 2010. "Studies show that a child who has been suspended is more likely to be retained in his or her grade, to drop out, to commit a crime, and to be incarcerated as an adult."[90]

Yet another trend was disheartening. At the same time that more children of color found themselves enmeshed already as preteens in the school-to-prison pipeline, there emerged strong statistical evidence that public school districts, already at the elementary level, were far more likely to exclude children from poorer backgrounds from placement in advanced classrooms—even when these children scored *better* on standardized tests than did their middle-class peers. This is what a comprehensive 2017 investigative study, conducted by two regional newspapers in North Carolina, found when it reviewed data for six years ending in 2015 from all 115 public school districts in the state—a survey of records for more than one million children each year. The North Carolina investigation determined that—over the course of its study—approximately nine thousand low-income elementary school children who scored higher than their middle-class peers on an end-of-year standardized math test were *not* tracked into gifted classrooms, while middle-income students with lower test scores were. The study further established that a failure to place poor children in these advanced classes had longer-term consequences. The North Carolina investigation found, for instance, that gifted low-income elementary school children were "less likely to take high school math in middle school, an important step toward the type of transcript that will open college doors." Even when low-income students did take middle-school math, they were still less likely than middle-class pupils with comparable test scores to take Advanced Placement math in high school. The study noted additionally that middle-class parents were far more willing (and financially able) to boost their children's test scores by seeking out private consultations with psychologists who openly advertised their capacity to arrange for children to receive an official label of "gifted-ness." The North Carolina investigation concluded, "The benefits of rigor-ous and challenging classes begin early and build over time. The effects are cumulative, since success in earlier grades leads to more opportunities and benefits in later years. Poor students with potential need help the most, and have the most to lose if they fall off the honors track."[91]

These findings raise a pressing question that the North Carolina report did not address. For if the much-touted "bell curve" of intelligence so con-sistently demonstrates a greater propensity for "gifted" talents among white children, as proponents keep insisting and claiming, why then would there be such a systematic scramble to suppress the educational potential of so many thousands of poor children of color? We are left only to wonder.

We may also be left to wonder what have been the long-term achieve-ments since *Brown v. Board of Education.* On the one hand, there has been genuine progress; the gap in test scores between white and African Ameri-can students in both reading and math has narrowed dramatically, and this convergence in academic achievement can be attributed principally to school desegregation and to "changes within schools."[92] On the other hand, there has been a decisive move to unravel *Brown* and to resegregate public educa-tion. It is a process that educator Jonathan Kozol already in 2005 caustically labeled "the restoration of apartheid schooling."[93]

This is a trend that now receives the blessings of the U.S. Supreme Court. Notably, for instance, the court in 2007 ruled in favor of white parents who sued to stop a city's plans to reduce racial imbalance on the grounds that it would be discriminatory against *white* children. Voting with the majority in *Parents Involved in Community Schools v. Seattle School District No. 1*, Chief Justice John Roberts, moreover, had the chutzpah to invoke *Brown*: "Before

Brown, schoolchildren were told where they could and could not go to school based on the color of their skin." Roberts added, "To the extent the objective is sufficient diversity so that students see fellow students as individuals rather than solely as members of a racial group, using means that treat students solely as members of a racial group is fundamentally at cross-purposes with that end." Roberts concluded, "The way to stop discrimination on the basis of race is to stop discrimination on the basis of race."[94] (To which Justice John Paul Stevens offered an especially acerbic dissent.)[95] Most recently, education analyst Richard Rothstein, author of *The Color of Law*, has pointed out that Chief Justice Roberts and his colleagues "got their facts wrong."[96] Nonetheless, since the 2007 *Parents Involved* decision, there has been not only a steady trend away from court-mandated orders to desegregate school districts but also a trend to prohibit school districts from implementing *voluntary* efforts to advance school integration. This double move appears to mark an end to an era of desegregation that had so momentously begun with *Brown* in 1954.[97]

In *The Bell Curve*, Herrnstein and Murray had admitted that theirs was a vision of a future for the United States that approximated the "apocalyptic." They acknowledged that they were deeply "pessimistic" about "the way we are headed" as a nation. "On the other hand," they emphasized, "there is much to be pessimistic about."[98] A similar pessimism, even an apocalyptic sense of despair — if not remotely for the same reasons that Herrnstein and Murray articulated — could apply to the lessons we can glean from studying a history of race and intelligence after *The Bell Curve*.

AFTERWORD

History often cycles its errors.

— Stephen Jay Gould, 1981

I N 1905 FRENCH PHYSIOLOGICAL psychologist Alfred Binet pioneered a "metrical scale of intelligence," a practical and easily administered system for establishing a child's capacity to perform complex mental processes. Binet did not intend his intelligence test — or the score that the test yielded — to be anything more than a method to identify, and thus to assist, children who experienced difficulties with learning. He meant his test above all to serve as a diagnostic — not a prognostic — tool. Binet himself doubted whether it was possible to come up with a mental test that might predict a child's potential for tackling a particular career or for developing prowess in specific life skills. He believed very much that a child's intelligence was malleable and could be improved with better educational interventions. It troubled him that intelligence testing might be misused in ways that would stigmatize lower-performing children. Soon thereafter (in 1912), however, German psychologist William Stern suggested a vital refinement of Binet's methods. Binet had aimed solely to determine a child's mental age; by contrast, Stern proposed that a child's cognitive performance be measured with a ratio that divided mental age by chronological age — and then multiplied by a factor of one hundred. It was Stern's equation that produced a single numerical index known as *intelligence quotient.*

When the concept of IQ arrived on American shores, it rapidly became racialized. An early adopter of IQ testing was Henry H. Goddard, who directed a school for "feeble-minded" youth in New Jersey. In 1913 Goddard successfully campaigned to conduct IQ tests on a relatively small cohort of immigrants as they arrived at New York City's Ellis Island; his results determined — perhaps unsurprisingly, given that many of his subjects were not fluent in English — that a remarkably high percentage of immigrants exhibited signs of mental deficiency. Not long afterward, Stanford psychologist Lewis M. Terman exponentially expanded the social and political ramifications of intelligence testing. Terman adapted and standardized Binet's methods with a test (which he named the Stanford-Binet in 1916) that could more easily be applied in several realms of public and professional life — including as a means to track children in schools, but also as a way to screen applicants for employment. Even more dramatically, Harvard psychologist Robert M. Yerkes, together with Goddard, persuaded the U.S. Army during the First World War to administer IQ tests to every army draftee (altogether 1.7 million men); the data (it was contended) clearly demonstrated the superior intelligence of men of northern European heritage over men of African descent. Princeton psychologist Carl Brigham, who had worked with Yerkes, promoted these results far more widely in a major study based on these army tests. (Brigham would — only a few years after this — devise the first standardized scholastic aptitude test, said accurately to predict a high school student's ability to succeed in college.) The army results soon thereafter came to serve as scientific evidence in propagandistic efforts to pass legislation severely to limit emigration from so-called undesirable regions around the world. Most devastating in this regard was the Johnson-Reed Immigration Act of 1924, which not only set harshly restrictive quotas on peoples from eastern and southern Europe but also banned all persons entering the United States from the Middle East and vast stretches of Asia (including Japan and India).

That the racialization of mental testing came so powerfully to thrive in the United States was due in no small part to the growing prestige and influence of the discipline of psychology. Not only Terman and Goddard but also Clark University professor G. Stanley Hall — arguably the nation's most prominent psychologist — actively promoted a theory of human intelligence that identified cognitive skills as fixed at birth. But there was more to it than that. Additionally, U.S. psychology came to imbibe the pernicious racialist views of British eugenicist Francis Galton. Already before Binet, Galton had come up with his own version of a mental test, even as he sought to use his test to demonstrate that differentials in intelligence between classes of people in a society reflected how natural the social order really was. According to Galton, the more innately intelligent a person was, the higher on the socioeconomic ladder that person inevitably turned out to be. And it was Galton's view that had leached by the early twentieth century into the groundwater of American psychology. American psychologists effectively turned the IQ test into a tool (or a weapon) to "prove" how inescapable differences in intelligence were between whites and blacks (not to mention between men and women and between wealthy and poor). With the best of confirmation biases, a "science" of mental testing came to identify the upper-class white man with his innately superior intelligence as the zenith of Western civilization. It would take the extreme logical extension of this position — the devastations of German Nazism and the atrocities of the Final Solution — to expose this shameful episode in the history of psychology for the sham that it had always been.

The story of race and intelligence did not end here. In the revised and expanded edition of his masterwork, *The Mismeasure of Man*, paleontologist and historian of science Stephen Jay Gould commented perceptively in his scathing critique of *The Bell Curve* how "innatist arguments for unitary, rankable intelligence" of the sort that Richard Herrnstein and Charles Murray so energetically promoted are "always present, always available, always published, always exploitable." Gould commented that resurgences in biological determinist thought consistently "correlate with episodes of political retrenchment, particularly with campaigns for reduced government spending on social programs, or at times of fear among ruling elites, when disadvantaged groups sow serious social unrest or even threaten to usurp power." Gould inquired rhetorically, "What argument against social change could be more chillingly effective than the claim that established orders, with some groups on top and others at the bottom, exist as an accurate reflection of the innate and unchangeable intellectual capacities of people so ranked?" And Gould went on to speculate how the "initial success" of *The Bell Curve* "in winning such attention must reflect the depressing temper of our time — a historical moment of unprecedented ungenerosity, when a mood for slashing social programs can be so abetted by an argument that beneficiaries cannot be aided due to inborn cognitive limits expressed as low IQ scores."[1] This remains a trenchant, compelling set of insights. Yet might there be another way to interpret the evidence from history?

It is no criticism of Gould to remark that his discussion of how biological determinist thought will always (and again) recur may have gotten the direction of causation switched around. Perhaps it is not — or may not *only* be — that the reappearance of these ungenerous and divisive theories of race and intelligence in American cultural and political life reflects the mood of a cultural moment when white people, of whatever class status, are experiencing spikes in their resentfulness toward the less fortunate in their midst.

It may instead be — or may *also* be — something of the reverse: that the articulations of these corrosive theories are purposive and productive and not just reflective; that they function much like a burning and blinding acid; that they can themselves stimulate and exacerbate cruel feelings; that these ideas can scar all persons of all backgrounds and races and classes in ways that it may not be possible entirely to undo or always to heal.

ACKNOWLEDGMENTS

I THANK THE FOLLOWING libraries and collections for indispensable assistance in the researching of this book: the Neilson Library and Young Science Library at Smith College, the Five College Library Repository Collection, the Firestone Library and Science Library at Princeton University, and the Lloyd Sealy Library at the John Jay College of Criminal Justice. I wish additionally to thank the provost's office at Baruch College and the Research Foundation of the City University of New York for financial support toward the preparation and completion of the manuscript. Sincerest thanks go as well to the wonderfully interdisciplinary Cheiron community of the International Society for the History of Behavioral and Social Sciences; Cheiron audiences provided excellent critiques of early versions of material included here. In particular, I offer appreciation to James Capshew, Cathy Faye, Ben Harris, Heather Murray, and Andrew Winston. For their willingness to comment (however critically) on earlier drafts of chapters, deepest thanks go to Mari Jo Buhle, Michelle Fine, Harry Heft, Jill Morawski, Alice O'Connor, Mike Pettit, Russ Poldrack, Hank Stam, Fernando Vidal, and Nadine Weidman. I extend a special thank you to Matthew Gambino and Mical Raz for their timely invitation to discuss my work in progress at the Yale School of Medicine — and to David Roediger for a crucial supportive letter out of the blue.

I am most thankful to all the individuals at the University of North Carolina Press for seeing this manuscript through to publication. Everyone at UNC Press has been immensely helpful, though no one more than my editor, Lucas Church. Lucas has been a paragon of efficiency; I am deeply indebted to him for his many encouragements along the way. I wish also to acknowledge Allan M. Brandt, Larry Churchill, and Jonathan Oberlander, the editors of the Studies in Social Medicine series. I am honored to be included on this list.

My students at Baruch College these last several years deserve a large amount of credit for being sources of inspiration (and welcome distractions) as I was working on this manuscript. I am infinitely grateful to them for their humor, humaneness, and often incidental gestures of kindness. Sincerest thanks as well go to the long list of people at the City University of New York who have provided me critical ballast in these difficult times. And I want to thank Tracey Wilson and Beth Bye for returning to my life with their graciousness and insight just when those qualities were most needed.

From initial ponderings to final edits, Dagmar Herzog made everything possible and everything better. I had no clue at the start what an unholy and complicated process writing and researching this project would turn out to be; any lesser soul would have urged me to toss in the sponge, as my mother liked to say. I cannot begin to express how appreciative I am that Dagmar chose to see this one through with me. My name is on the book's spine, but whatever is good and right and true and lasting in this volume belongs entirely to her. The flaws herein remain all mine.

NOTES

Introduction

1. Chief Justice Earl Warren's opinion in *Brown v. Board of Education of Topeka* (1954), as reprinted in Richard Kluger, *Simple Justice: The History of "Brown v. Board of Education" and Black America's Struggle for Equality* (New York: Knopf, 1976), 781.

2. Daryl Michael Scott, *Contempt and Pity: Social Policy and the Image of the Damaged Black Psyche, 1880–1996* (Chapel Hill: University of North Carolina Press, 1997), 120. Scott added that it has remained a common misreading of *Plessy* to posit that it addressed the psychological damage — rather than the social consequences — of state-sponsored segregation. Scott argued that what was at stake in *Plessy* "was whether segregation damaged a black person's reputation, not his psyche" (119–20).

3. Justice Henry Brown, "Majority Opinion in *Plessy v. Ferguson*, 1896," in Waldo E. Martin Jr., *"Brown v. Board of Education": A Brief History with Documents* (Boston: Bedford/St. Martin's, 1998), 80.

4. Tourgée then inquired, "Indeed, is [whiteness] not the most valuable sort of property, being the master-key that unlocks the golden door of opportunity?" Cited in Albert P. Blaustein and Robert L. Zangrando, *Civil Rights and African Americans: A Documentary History* (Evanston: Northwestern University Press, 1968), 300. Also see the discussion of *Plessy* in Cheryl I. Harris, "Whiteness as Property," *Harvard Law Review* 106 (June 1993): esp. 1746–50.

5. The *Plessy* decision declared, "If he be a white man, and assigned to a colored coach, he may have his action for damages against the company for being deprived of his so-called 'property.' Upon the other hand, if he be a colored man, and be so assigned, he has been deprived of no property, since he is not lawfully entitled to the reputation of being a white man." Quoted in Harris, "Whiteness as Property," 1749.

6. See Warren's opinion in *Brown*, as reprinted in Kluger, *Simple Justice*, 781. Also see the analysis of legal historian Risa L. Goluboff, who has argued that "the focus on both stigma and state action in *Brown* subordinated the material to the psychological. In doing so, it subordinated the problems most acute for working African Americans to those most acute for the more privileged of the race." Risa L. Goluboff, *The Lost Promise of Civil Rights* (Cambridge, Mass.: Harvard University Press, 2007), 252.

7. In addition to *Brown* in Kansas and *Briggs v. Elliott* in South Carolina, the school desegregation cases (all represented by the NAACP Legal Defense Fund) were from Virginia (*Davis v. County School Board of Prince Edward County*) and Delaware (*Gebhart v. Belton*).

8. See, for instance, Kenneth B. Clark and Mamie P. Clark, "Racial Identification and Preference in Negro Children," in *Readings in Social Psychology*, ed. Theodore Newcomb and Eugene Hartley (New York: Henry Holt, 1947), 169–78; and Kenneth B. Clark and Mamie P. Clark, "Emotional Factors in Racial Identification and Preference in Negro Children," *Journal of Negro Education* 19 (Summer 1950): 341–50. Also see Gwen Bergner, "Black Children, White Preference: *Brown v. Board*, the Doll Tests, and the Politics of Self-Esteem," *American Quarterly* 61 (June 2009): 299–332.

9. "Testimony of Kenneth Clark — *Briggs* Trial," in *Removing a Badge of Slavery: The Record of*

"Brown v. Board of Education," ed. Mark Whitman (New York: Markus Wiener, 1993), 49–51. On the NAACP Legal Defense Fund strategy in the *Briggs* case to foreground the psychological damage of segregation, see Idus A. Newby, *Challenge to the Court: Social Scientists and the Defense of Segregation, 1954–1966* (Baton Rouge: Louisiana State University Press, 1969), 19–42.

10. Ellen Herman, *The Romance of American Psychology: Political Culture in the Age of Experts* (Berkeley: University of California Press, 1995), 195–96.

11. See Scott, *Contempt and Pity*, esp. 71–136. For a more sympathetic interpretation of Kenneth Clark's research on racial preference and its impact on postwar U.S. culture and politics, see Ben Keppel, *The Work of Democracy: Ralph Bunche, Kenneth B. Clark, Lorraine Hansberry, and the Cultural Politics of Race* (Cambridge, Mass.: Harvard University Press, 1995), esp. 97–131.

12. John P. Jackson Jr., "The Scientific Attack on *Brown v. Board of Education*, 1954–1964," *American Psychologist* 59 (September 2004): 531.

13. The testimony of psychologist R. Travis Osborne in *Stell v. Savannah-Chatham County Board of Education* as cited by the Fifth Circuit Court of Appeals in its 1965 decision. See 255 F. Supp. 83 (S.D. Ga. 1965), available at https://casetext.com/case/stell-v-savannah-chatham-county-board -of-education-2 (accessed March 12, 2018). Also see R. T. Osborne, "Racial Differences in Mental Growth and School Achievement: A Longitudinal Study," *Psychological Reports* 7 (October 1960): 233–39.

14. Judge Frank M. Scarlett's *Stell* opinion (1963) quoted in Richard R. Valencia, *Dismantling Contemporary Deficit Thinking: Educational Thought and Practice* (New York: Routledge, 2010), 27.

15. William H. Tucker, *The Science and Politics of Racial Research* (Urbana: University of Illinois Press, 1994), 166.

16. For histories of the *Stell* case, see Newby, *Challenge to the Court*, 191–210; and John P. Jackson Jr., *Science for Segregation: Race, Law, and the Case against "Brown v. Board of Education"* (New York: New York University Press, 2005), 131–47.

17. Henry E. Garrett, "The Equalitarian Dogma," *Perspectives in Biology and Medicine* 4 (Summer 1961): 480, 484.

18. Audrey M. Shuey, *The Testing of Negro Intelligence*, 2nd ed. (New York: Social Science Press, 1966), 520–21. For a contemporaneous rebuttal of Shuey and "the 'scientific racist' position," see Thomas Pettigrew, "Negro American Intelligence: A New Look at an Old Controversy," *Journal of Negro Education* 33 (Winter 1964): 6–25. For the view that Shuey's book "rekindled the coals leading to a revival of modern day hereditarianism," see Richard R. Valencia and Daniel G. Solórzano, "Contemporary Deficit Thinking," in *The Evolution of Deficit Thinking: Educational Thought and Practice*, ed. Richard R. Valencia (Washington, D.C.: Falmer Press, 1997), 161.

19. Sherwood Washburn, review of *Race and Modern Science*, edited by Robert E. Kuttner, in *American Anthropologist* 70 (October 1968): 1036. Also see Robert E. Kuttner, ed., *Race and Modern Science* (New York: Social Science Press, 1967).

20. Arthur R. Jensen, "How Much Can We Boost IQ and Scholastic Achievement?," *Harvard Educational Review* 39 (Winter 1969): 82–83. A history of scientific racism observed, "Immediately upon publication, Jensen's argument became more famous than nearly any other psychological paper of the twentieth century." John P. Jackson Jr. and Nadine M. Weidman, *Race, Racism, and Science: Social Impact and Interaction* (Denver: ABC-CLIO, 2004), 221.

21. Richard J. Herrnstein, "I.Q.," *Atlantic*, September 1971, 58. Also see R. J. Herrnstein, *I.Q. in the Meritocracy* (Boston: Little, Brown, 1973).

22. The seemingly color-blind talk and the focus on the inside of individuals' brains coexist with devastating long-term trends that keep substantial portions of the population permanently disempowered and in poverty. One important recent study demonstrates the long-term patterns in housing segregation enforced by U.S. law (and not just by custom) — and the repeated restriction of individuals of color to poorer and more dangerous and unhealthy neighborhoods. See Richard

Rothstein, *The Color of Law: A Forgotten History of How Our Government Segregated America* (New York: Liveright, 2017). To take another key example: Beyond the scope of the narrative of this book, but closely related to it, is the evolution of mass incarceration in the United States. A number of recent studies have pointed out, moreover, that the rapid expansion of the carceral state coincided with more liberal programs of the War on Poverty and was also promoted by Democrats. For instance, see Naomi Murakawa, *The First Civil Right: How Liberals Built Prison America* (New York: Oxford University Press, 2014); and Elizabeth Hinton, *From the War on Poverty to the War on Crime: The Making of Mass Incarceration in America* (Cambridge, Mass.: Harvard University Press, 2016). For an argument that the origins of mass incarceration in the United States can be traced much further back — at least to the beginning of the twentieth century — see Khalil Gibran Muhammad, *The Condemnation of Blackness: Race, Crime, and the Making of Modern Urban America* (Cambridge, Mass.: Harvard University Press, 2011).

23. For one study among many of the larger onslaught, see Diane Ravitch, *Reign of Error: The Hoax of the Privatization Movement and the Danger to America's Public Schools* (New York: Knopf, 2013).

24. Mary L. Dudziak, "Desegregation as a Cold War Imperative," *Stanford Law Review* 61 (November 1988): 61–120.

25. Mark Tushnet with Katya Lezin, "What Really Happened in *Brown v. Board of Education*," *Columbia Law Review* 91 (December 1991): 1867–1930. A. Philip Randolph of the Brotherhood of Sleeping Car Porters triumphantly declared that the *Brown* decision "will effectively impress upon millions of colored people in Asia and Africa the fact that idealism and social morality can and do prevail in the United States, regardless of race, creed, or color." Quoted in James T. Patterson, *"Brown v. Board of Education": A Civil Rights Milestone and Its Troubled Legacy* (New York: Oxford University Press, 2001), 71.

26. See Christopher S. Parker, *Fighting for Democracy: Black Veterans and the Struggle against White Supremacy in the Postwar South* (Princeton: Princeton University Press, 2009).

27. Frances Fox Piven and Richard A. Cloward, *Regulating the Poor: The Functions of Public Welfare* (New York: Pantheon, 1971), 229–30.

28. Derrick A. Bell Jr., "*Brown v. Board of Education* and the Interest-Convergence Dilemma," *Harvard Law Review* 93 (January 1980): 523–25.

29. Derrick A. Bell, *Silent Covenants: "Brown v. Board of Education" and the Unfulfilled Hopes for Racial Reform* (New York: Oxford University Press, 2004), 59. Noting that "a fortuitous symmetry existed between what blacks sought and what the nation needed," Bell added, "I do not intend by this conclusion to belittle the NAACP lawyers' long years of hard work and their carefully planned strategies that brought the cases consolidated in *Brown v. Board of Education* to the Supreme Court" (59–60).

30. Michelle Alexander, *The New Jim Crow: Mass Incarceration in the Age of Colorblindness* (New York: New Press, 2010), 256.

31. Michael J. Klarman, "How *Brown* Changed Race Relations: The Backlash Thesis," *Journal of American History* 81(June 1994): 82.

32. Lani Guinier, "From Racial Liberalism to Racial Literacy: *Brown v. Board of Education* and the Interest-Divergence Dilemma," *Journal of American History* 91 (June 2004): 102. Guinier added that *Brown* "intensified divergences between northern elites and southern whites, solidified the false interest convergence between southern white elites and southern poor whites, ignored the interest divergences between poor and middle-class blacks, and exacerbated the interest divergences between poor and working-class whites and blacks" (102).

33. See Arlie Russell Hochschild, "Appendix B: Politics and Pollution" and "Appendix C: Fact-Checking Common Impressions," in *Strangers in Their Own Land: Anger and Mourning on the American Right* (New York: New Press, 2016), 251–61.

Chapter 1

1. "Excerpts from Nixon's Poverty Message," *New York Times*, February 20, 1969, 33.

2. Richard Nixon, "Statement Announcing the Establishment of the Office of Child Development," April 9, 1969, *Public Papers of the Presidents of the United States: Richard Nixon: Containing the Public Messages, Speeches, and Statements of the President, 1969* (Washington, D.C.: U.S. Government Printing Office, 1971), 270–71.

3. For a contemporaneous account of this history, see Walter Williams and John W. Evans, "The Politics of Evaluation: The Case of Head Start," *Annals of the American Academy of Political and Social Science* 385 (September 1969): 118–32. On efforts by the Nixon administration to defund Head Start by casting the program as a failure, see Alice O'Connor, *Poverty Knowledge: Social Science, Social Policy, and the Poor in Twentieth-Century U.S. History* (Princeton: Princeton University Press, 2001), esp. 187–88.

4. On Head Start as a predominantly African American program, see Jill S. Quadagno, *The Color of Welfare: How Racism Undermined the War on Poverty* (New York: Oxford University Press, 1994), 142.

5. An idea that the racial bias that saturated American culture might well be the primary cause for the lower IQ rankings of African Americans had existed at least since psychologist Donald O. Hebb posed the question, "Negroes living in the United States make lower average scores on intelligence tests than whites do, but we cannot conclude that the Negro has a poorer brain than the white. Why?" Hebb continued: "Because Negro and white do not have the opportunity to learn to speak the language with equal range and accuracy, are not usually taught in equally good schools, and do not have equally good jobs or equal exposure to cultural influences that usually require a fairly good income." Donald O. Hebb, *The Organization of Behavior: A Neuropsychological Theory* (New York: Wiley, 1949), 300. Also see Maris A. Vinovskis, *The Birth of Head Start: Preschool Education Policies in the Kennedy and Johnson Administrations* (Chicago: University of Chicago Press, 2005), 9–10.

6. Joseph McVicker Hunt, *Intelligence and Experience* (New York: Ronald, 1961), 363.

7. Joseph McVicker Hunt, "The Psychological Basis for Using Pre-school Enrichment as an Antidote for Cultural Deprivation," *Merrill-Palmer Quarterly of Behavior and Development* 10 (July 1964): 210, 236.

8. David P. Ausubel, "A Teaching Strategy for Culturally Deprived Pupils: Cognitive and Motivational Considerations," *School Review* 71 (Winter 1963): 454. Also see David P. Ausubel, "How Reversible Are the Cognitive and Motivational Effects of Cultural Deprivation? Implications for Teaching the Culturally Deprived Child," *Urban Education* 1 (Summer 1964): 16–38.

9. Benjamin S. Bloom, *Stability and Change in Human Characteristics* (New York: Wiley and Sons, 1964), 88. Bloom added, "These results make it clear that a single early measure of general intelligence cannot be the basis for a long-term decision about an individual" (88).

10. Benjamin S. Bloom, Allison Davis, and Robert Hess, *Compensatory Education for Cultural Deprivation* (New York: Holt, Rinehart, and Winston, 1965), 15. Also see Jerome S. Bruner, *The Process of Education* (Cambridge, Mass.: Harvard University Press, 1960).

11. Mical Raz, *What's Wrong with the Poor? Psychiatry, Race, and the War on Poverty* (Chapel Hill: University of North Carolina Press, 2013), 63. Also see Harold Silver and Pamela Silver, *An Educational War on Poverty: American and British Policy-Making, 1960–1980* (New York: Cambridge University Press, 1991).

12. Bettye M. Caldwell, "The Rationale for Early Intervention," *Exceptional Children* 36 (Summer 1970): 721.

13. On Seligman and positive psychology, see Martin E. P. Seligman, *Learned Optimism* (New York: Knopf, 1991). There are several excellent sources that detail how psychologists John Bruce Jessen and James Mitchell "reverse-engineered" Seligman's 1960s learned helplessness experiments

with dogs to design CIA torture methods. These include Jane Mayer, "The Experiment," *New Yorker*, July 11, 2005; Katherine Eban, "The War on Terror: Rorschach and Awe," *Vanity Fair*, July 2007; Jane Mayer, *The Dark Side: The Inside Story of How the War on Terror Turned into a War on American Ideals* (New York: Doubleday, 2008); Stephen Soldz, "Psychologists, Torture, and Civil Society: Complicity, Institutional Failure, and the Struggle for Professional Transformation," in *The United States and Torture: Interrogation, Incarceration, and Abuse*, ed. Marjorie Cohn (New York: New York University Press, 2011), 177–202; James Risen, *Pay Any Price: Greed, Power, and Endless War* (New York: Houghton Mifflin Harcourt, 2014), esp. 177–201; and Tamsin Shaw, "The Psychologists Take Power," *New York Review of Books*, February 25, 2016. For psychologists Jonathan Haidt and Steven Pinker's impassioned rebuttal to Shaw's charge that Seligman collaborated with Jessen and Mitchell, see "Moral Psychology: An Exchange," *New York Review of Books*, April 7, 2016.

14. See, for example, J. Bruce Overmier and Martin E. P. Seligman, "Effects of Inescapable Shock upon Subsequent Escape and Avoidance Responding," *Journal of Comparative and Physiological Psychology* 63 (February 1967): 28–33; and Martin E. P. Seligman and Steven F. Maier, "Failure to Escape Traumatic Shock," *Journal of Experimental Psychology* 74 (May 1967): 1–9.

15. Martin E. P. Seligman, "For Helplessness: Can We Immunize the Weak?," *Psychology Today*, June 1969, 42–44.

16. For a learned helplessness study conducted with albino rats, see Martin E. P. Seligman, "Chronic Fear Produced by Unpredictable Electric Shock," *Journal of Comparative and Physiological Psychology* 66 (October 1968): 402–11. For a study conducted with mongrel dogs, see Martin E. P. Seligman, Steven F. Maier, and James H. Geer, "Alleviation of Learned Helplessness in the Dog," *Journal of Abnormal Psychology* 73 (June 1968): 256–62.

17. Quoted in Robert Rosenthal and Lenore Jacobson, *Pygmalion in the Classroom: Teacher Expectation and Pupils' Intellectual Development* (New York: Holt, Rinehart and Winston, 1968), 183.

18. Robert Rosenthal and Kermit L. Fode, "The Effect of Experimenter Bias on the Performance of the Albino Rat," *Behavioral Science* 8 (July 1963): 184. Also see Robert Rosenthal and Reed Lawson, "A Longitudinal Study of the Effects of Experimenter Bias on the Operant Learning of Laboratory Rats," *Journal of Psychiatric Research* 2 (June 1964): 61–72.

19. Robert Rosenthal, "On the Social Psychology of the Psychology Experiment: The Experimenter's Hypothesis as Unintended Determinant of Experimental Results," *American Scientist* 51 (June 1963): 269.

20. Rosenthal and Fode, "Effect of Experimenter Bias," 184.

21. Rosenthal and Jacobson, *Pygmalion in the Classroom*, 38.

22. See Robert Rosenthal, *Experimenter Effects in Behavioral Research* (New York: Appleton-Century-Crofts, 1966). Also see the experiment conducted by psychologist David Rosenhan (at almost the same moment Rosenthal was deceiving his students about "maze-bright" rats): Rosenhan informed his research assistant that he wished to replicate the results of a prior study. Rosenhan provided his assistant with results that were, in fact, *the direct opposite* of the results he had actually obtained in that prior study. The research assistant ran the identical trial again in an identical manner. She was able to collect data that supported not the original results Rosenhan had obtained but rather the falsified outcome Rosenhan had presented to her beforehand and that she had been led to anticipate. Like Rosenthal, Rosenhan argued that psychological experimentation was a self-fulfilling prophecy. See David L. Rosenhan, "On the Social Psychology of Hypnosis Research," *Education Testing Services Research Memorandum*, March 1964.

23. Rosenthal, "On the Social Psychology of the Psychology Experiment," 269, 280. Already in 1962, at the conference "Education in Depressed Areas," held at the Teachers College, Columbia University, the director of the Great Cities Program for School Improvement, Detroit Public Schools, Carl L. Marburger, cited Rosenthal's research with rats before he announced, "If nothing else, teacher expectations affect the time spent in preparing to teach, the amount of real concern for individual students, and the degree of 'soul' the teacher gives to his work." See Carl L. Marburger,

"Considerations for Educational Planning," in *Education in Depressed Areas*, ed. A. Harry Passow (New York: Teachers College Press, 1963), 307.

24. Robert Rosenthal and Lenore Jacobson, "Pygmalion in the Classroom," *Urban Review* 3 (September 1968): 16.

25. Robert Rosenthal and Lenore Jacobson, "Teachers' Expectancies: Determinants of Pupils' IQ Gains," *Psychological Reports* 19 (August 1966): 116.

26. Robert Rosenthal and Lenore F. Jacobson, "Teacher Expectations for the Disadvantaged," *Scientific American*, April 1968, 22.

27. Rosenthal and Jacobson, *Pygmalion in the Classroom*, 135.

28. For an introduction to implicit bias, see Mahzarin R. Banaji and Anthony G. Greenwald, *Blindspot: Hidden Biases of Good People* (New York: Delacorte Press, 2013). This text does not acknowledge the historical precedent of the Pygmalion effect.

29. J. Michael Palardy, "What Teachers Believe—What Children Achieve," *Elementary School Journal* 69 (April 1969): 370–74.

30. "Teachers: Blooming by Deception," *Time*, September 20, 1968, 62.

31. N. L. Gage, "Preface," in *Pygmalion Reconsidered: A Case Study in Statistical Inference: Reconsideration of the Rosenthal-Jacobson Data on Teacher Expectancy*, ed. Janet D. Elashoff and Richard E. Snow (Worthington, Ohio: Charles A. Jones Publishing, 1971), v.

32. Roger Brown, *Social Psychology*, 2nd ed. (New York: Free Press, 1986), 517.

33. John Holt, *How Children Fail* (New York: Pitman, 1964), 58–59.

34. Charles E. Silberman, *Crisis in Black and White* (New York: Random House, 1964), 262.

35. Benjamin M. Spock, "Children and Discrimination," *Integrated Education* 2 (February 1964): 9.

36. Kenneth B. Clark, "The Cult of Cultural Deprivation: A Complex Social Psychological Phenomenon," in *Children with Reading Problems: Classic and Contemporary Issues in Reading Disability*, ed. Gladys Natchez (New York: Basic Books, 1968), 183. Clark first presented this talk at the conference "Environmental Deprivation and Enrichment: Proceedings of the Annual Invitational Conference on Urban Education" at Yeshiva University on April 26, 1965.

37. The story appears both in Kenneth B. Clark, "Educational Stimulation of Racially Disadvantaged Children," in Passow, *Education in Depressed Areas*, 148; and in Silberman, *Crisis in Black and White*, 261. Silberman attributes the story to Frank Riessman's *The Culturally Deprived Child* (New York: Harper and Row, 1962).

38. Arthur Pearl, "Grouping Hurts the Poor," *Southern Education Report* 2 (December 1966): 8

39. U.S. Commission on Civil Rights, *Racial Isolation in the Public Schools* (Washington, D.C.: Government Printing Office, 1967), 124. Although as Diane Ravitch has emphasized, "The report's judgments about the value of compensatory education in Negro schools were devastatingly negative," as it underscored that "the problems they attempted to solve stemmed in large measure 'from racial and social class isolation in schools which themselves are isolated by race and social class.'" Diane Ravitch, *The Troubled Crusade: American Education, 1945–1980* (New York: Basic Books, 1983), 172.

40. Cited in Alex Poinsett, "Ghetto Schools: An Educational Wasteland," *Ebony* 22 (August 1967): 54–55.

41. Quoted in Fred M. Hechinger, "The Teacher Gets What He Expects," *New York Times*, August 13, 1967, E9. Also see Robert Rosenthal and Lenore Jacobson, "Self-Fulfilling Prophecies in the Classroom: Teachers' Expectations as Unintended Determinants of Pupils' Intellectual Competencies," in *Social Class, Race, and Psychological Development*, ed. Martin Deutsch, Irwin Katz, and Arthur R. Jensen (New York: Holt, Rinehart and Winston, 1968), 219–53.

42. Daniel Schreiber quoted in U.S. Commission on Civil Rights, *Racial Isolation in the Public Schools*, 123.

43. U.S. Commission on Civil Rights, *Racial Isolation in the Public Schools*, 123.

44. Dr. Samuel Shepard quoted in Robert H. Collins, "Motivation," *Southern Education Report* 1 (July–August 1965), 24.

45. Collins, "Motivation," 26.

46. Kenneth B. Clark, *Dark Ghetto: Dilemmas of Social Power* (New York: Harper and Row, 1965), 144.

47. Quoted in John Holt, *How Children Fail*, rev. ed. (New York: Delta/Seymour Lawrence, 1982), 287.

48. Julian B. Rotter, *Social Learning and Clinical Psychology* (New York: Prentice-Hall, 1954), 120.

49. Julian B. Rotter, "Generalized Expectancies for Internal versus External Control of Reinforcement," *Psychological Monographs: General and Applied* 80 (1966): 1, 23.

50. Esther S. Battle and Julian B. Rotter, "Children's Feelings of Personal Control as Related to Social Class and Ethnic Group," *Journal of Personality* 31 (December 1963): 484–85, 488–89.

51. Pearl Mayo Gore and Julian B. Rotter, "A Personality Correlate of Social Action," *Journal of Personality* 31 (March 1963): 58–64.

52. Bonnie Ruth Strickland, "The Prediction of Social Action from a Dimension of Internal-External Control," *Journal of Social Psychology* 66 (August 1965): 357.

53. Herbert M. Lefcourt and Gordon W. Ladwig, "The American Negro: A Problem in Expectancies," *Journal of Personality and Social Psychology* 1 (April 1965): 378–80.

54. Herbert M. Lefcourt, "Risk Taking in Negro and White Adults," *Journal of Personality and Social Psychology* 2 (November 1965): 765, 769.

55. Quoted in James S. Coleman et al., *Equality of Educational Opportunity* (Washington, D.C.: U.S. Government Printing Office, 1966), iii.

56. Frederick Mosteller and Daniel P. Moynihan, eds., *On Equality of Educational Opportunity* (New York: Random House, 1972), dust jacket. Mosteller and Moynihan wrote that the findings of the Coleman Report "constitute the most powerful empirical critique of the myths (the unquestioned basic assumptions, the socially received beliefs) of American education ever produced. It is the most important source of data on the sociology of American education yet to appear." Mosteller and Moynihan, "A Pathbreaking Report," in Mosteller and Moynihan, *On Equality of Educational Opportunity*, 5.

57. Carlos F. Diaz, "Coleman Report," in *Encyclopedia of Diversity in Education*, vol. 1, ed. James A. Banks (Washington, D.C.: SAGE, 2012), 402. Diaz's summary of the Coleman Report does not mention the report's reliance on expectancy research for its analysis. For an influential critique of the report's methodological weaknesses, see Ronald P. Carver, "The Coleman Report: Using Inappropriately Designed Achievement Tests," *American Educational Research Association Journal* 12 (Winter 1975): 77–86.

58. David A. Gamson, Kathryn A. McDermott, and Douglas S. Reed, "The Elementary and Secondary Education Act at Fifty: Aspirations, Effects, and Limitations," *Russell Sage Foundation Journal of the Social Sciences* 1 (December 2015): 1.

59. Lyndon B. Johnson quoted in Charles Mohr, "President Signs Education Bill at His Old School," *New York Times*, April 12, 1965, 22.

60. Gerald Grant, "Shaping Social Policy: The Politics of the Coleman Report," *Teachers College Record* 75 (September 1973): 41.

61. Johnson quoted in Erwin Knoll, "Project Head Start Limps Along to Popularity," *Southern Education Report* 2 (July–August 1966): 8.

62. For example, see Brent Staples, "Where Did All the Black Teachers Go?," *New York Times*, April 20, 2017, A22. For the Coleman Report's own views on African American teachers, see Coleman et al., *Equality of Educational Opportunity*, 341, 344.

63. James S. Coleman quoted in Jim Leeson, "Questions, Controversies and Opportunities," *Southern Education Report* 1 (November–December 1965): 7.

64. Coleman et al., *Equality of Educational Opportunity*, 218, 325.

65. Ravitch, *Troubled Crusade*, 169.

66. Coleman et al., *Equality of Educational Opportunity*, 202, 320–21.

67. James S. Coleman, *Resources for Social Change: Race in the United States* (New York: Wiley-Interscience, 1971), 28.

68. Daniel Patrick Moynihan, *The Negro Family: The Case for National Action* (Washington, D.C.: Office of Policy Planning and Research, U.S. Department of Labor, March 1965), 25.

69. Daniel P. Moynihan, "Sources of Resistance to the Coleman Report," *Harvard Educational Review* 38 (Winter 1968): 33. Moynihan gave ample space in a collection he coedited to the view that the Coleman Report proved that "there does not seem to be any way for blacks to catch up with whites if family factors are ignored." See David J. Armor, "School and Family Effects on Black and White Achievement: A Reexamination of the USOE Data," in Mosteller and Moynihan, *On Equality of Educational Opportunity*, 225. After Coleman's death in 1995, Moynihan recounted an anecdote in which political sociologist Seymour Martin Lipset told him already in 1966 that the Coleman Report demonstrated how a disadvantaged child's achievement issues were "all family." See Daniel P. Moynihan, "James S. Coleman: Moved by the Data, Not Doctrine," *New York Times Magazine*, December 31, 1995, 25.

70. John Ehrlichman quoted in Robert B. Semple Jr., "The Middle American Who Edits Ideas for Nixon," *New York Times Magazine*, April 12, 1970, 74.

71. "Text of the President's Statement Explaining His Policy on School Desegregation," *New York Times*, March 25, 1970, 26.

72. Matthew D. Lassiter, *The Silent Majority: Suburban Politics in the Sunbelt South* (Princeton: Princeton University Press, 2006), 308.

73. See Lawrence J. McAndrews, *The Era of Education: The Presidents and the Schools, 1965–2001* (Urbana: University of Illinois Press, 2006), 64.

74. Alexander M. Bickel, "Desegregation: Where Do We Go from Here?," *New Republic*, February 7, 1970, 22. Also see Alexander M. Bickel, "'Realistic, Sensible,'" *New Republic*, April 4, 1970, 14–15.

75. Grant, "Shaping Social Policy," 24.

76. Coleman quoted in Jack Rosenthal, "School Expert Calls Integration Vital Aid to Educating the Disadvantaged," *New York Times*, March 9, 1970, 1.

77. Testimony of James Coleman before the Select Committee on Equal Educational Opportunity (April 21, 1970), in *Equal Educational Opportunity: Hearings* (Washington, D.C.: U.S. Government Printing Office, 1970), 93.

78. Thomas Pettigrew, *The Consequences of Racial Isolation in the Public Schools—Another Look* (Washington, D.C.: U.S. Department of Health, Education, and Welfare, 1967), 21–22.

79. J. McVicker Hunt, "Poverty versus Equality of Opportunity" (1967), reprinted in Hunt, *The Challenge of Incompetence and Poverty: Papers on the Role of Early Education* (Urbana: University of Illinois Press, 1969), 213–14, 216.

80. Arthur R. Jensen, "How Much Can We Boost IQ and Scholastic Achievement?," *Harvard Educational Review* 39 (Winter 1969): 84–86.

81. Fred M. Hechinger, "Curtain for Higher Horizons," *New York Times*, July 10, 1966, E7. It is worth noting that a defense of Higher Horizons was that this program did not fail so much as — due to its rapid growth — it began to suffer from a "complete dilution of services." See Elliott Shapiro, "A Child's Right to Childhood," *Integrated Education* 5 (October–November 1967): 34.

82. Edmund W. Gordon and Adelaide Jablonsky, "Compensatory Education in the Equalization of Educational Opportunity, I," *Journal of Negro Education* 37 (Summer 1968): 269, 271.

83. Westinghouse Learning Corporation, *The Impact of Head Start: An Evaluation of the Effects of Head Start on Children's Cognitive and Affective Development* (Washington, D.C.: Office of Economic Opportunity, 1969), 9. Policy analyst David Kirp would pointedly remark several decades later, "Never mind that the Westinghouse study was a flawed piece of research, subsequently picked to

death by other social scientists. Never mind, either, that it mainly focused on the summer program, a hastily mounted and soon abandoned effort, rather than the school-year [Head Start] program for three- and four-year-olds." David L. Kirp, *The Sandbox Investment: The Preschool Movement and Kids-First Politics* (Cambridge, Mass.: Harvard University Press, 2007), 62.

84. In 1972 the journal *Education and Urban Society* did conclude that "unfortunately the Coleman Report, with data collected on a national basis, does not support the generalization that desegregation will automatically lead to an increase in school achievement for the black and brown students." See Mark R. Lohman, "Changing a Racial Status Ordering: Some Implications for Integration Efforts," *Education and Urban Society* 4 (August 1972): 400.

85. Deborah W. Meier, "The Coleman Report," *Integrated Education* 5 (December 1967–January 1968): 37, 43–44. For a critique of the report's conclusion "that per-pupil expenditure and school facilities show very little relation to student achievement levels," also see Samuel Bowles and Henry M. Levin, "The Determinants of Scholastic Achievement: An Appraisal of Some Recent Evidence," *Journal of Human Resources* 3 (Winter 1968): 3–24.

86. See John Leo, "Study Indicates Pupils Do Well When Teacher Is Told They Will," *New York Times*, August 8, 1967. Also see Herbert R. Kohl, "Great Expectations," *New York Review of Books*, September 12, 1968, 30–31; and Robert Coles, "What Can You Expect?," *New Yorker*, April 19, 1969, 169–76.

87. Coleman et al., *Equality of Educational Opportunity*, 207.

88. See Samuel S. Wineburg, "The Self-Fulfillment of the Self-Fulfilling Prophecy," *Educational Researcher* 16 (December 1987): 32–33.

89. Len Sperry, "Community Control: A Proposal," *Elementary School Journal* 73 (December 1972): 144. Rosenthal and Jacobson's evidence was also cited in support of community control in Mario Fantini, Marilyn Gittell, and Richard Magat, *Community Control and the Urban School* (New York: Praeger, 1970).

90. Janet D. Elashoff and Richard E. Snow, "Pygmalion in the Classroom as a Report of Original Research," in Elashoff and Snow, *Pygmalion Reconsidered*, 9

91. Rosenthal and Jacobson, *Pygmalion in the Classroom*, 50–51.

92. Rosenthal and Jacobson, "Teacher Expectations for the Disadvantaged," 19, 23.

93. Robert L. Thorndike, review of *Pygmalion in the Classroom*, by Robert Rosenthal and Lenore Jacobson, *American Educational Research Journal* 5 (November 1968): 708.

94. Lee J. Cronbach, "Five Decades of Public Controversy over Mental Testing," in *Controversies and Decisions: The Social Sciences and Public Policy*, ed. Charles Frankel (New York: Russell Sage Foundation, 1976), 134.

95. John P. Spencer, *In the Crossfire: Marcus Foster and the Troubled History of American School Reform* (Philadelphia: University of Pennsylvania Press, 2012), 260.

96. William L. Claiborn, "Expectancy Effects in the Classroom: A Failure to Replicate," *Journal of Educational Psychology* 60 (1969): 377–83. When another study also failed to replicate Rosenthal and Jacobson's results, its conclusions were blunt: "No significant differences were found in IQ, achievement, students' grades or behavior, and no differences were observed in teacher behavior." See Jean José and John J. Cody, "Teacher-Pupil Interaction as It Relates to Attempted Changes in Teacher Expectancy of Academic Ability and Achievement," *American Educational Research Journal* 8 (January 1971): 47.

97. Barbara Beatty, "The Debate over the Young 'Disadvantaged Child': Preschool Intervention, Developmental Psychology, and Compensatory Education in the 1960s and Early 1970s," *Teachers College Record* 114 (June 2012): 22. While she discusses Jensen's critiques, Beatty does not mention *Pygmalion in the Classroom* or how the book intersected with contemporary debates over intelligence.

98. Jensen, "How Much Can We Boost IQ and Scholastic Achievement?," 29, 41.

99. Arthur R. Jensen, review of *Pygmalion in the Classroom*, by Robert Rosenthal and Lenore Jacobson, *American Scientist* 57 (Spring 1969): 45A.

100. See Russell C. Leaf, "Avoidance Response Evocation as a Function of Prior Discriminative Fear Conditioning under Curare," *Journal of Comparative and Physiological Psychology* 58 (December 1964): 446–49; and Bruce Overmier and Russell C. Leaf, "Effects of Discriminative Pavlovian Fear Conditioning upon Previously or Subsequently Acquired Avoidance Responding," *Journal of Comparative and Physiological Psychology* 60 (October 1965): 213–17. Also see the discussion of these early experiments in Christopher Peterson, Steven F. Maier, and Martin E. P. Seligman, *Learned Helplessness: A Theory for the Age of Personal Control* (New York: Oxford University Press, 1993), 17–20.

101. Steven F. Maier and Martin E. P. Seligman, "Learned Helplessness at Fifty: Insights from Neuroscience," *Psychological Review* 123 (July 2016): 349.

102. Seligman, "For Helplessness," 43. It took dragging the dogs to safety *more than thirty times* before these dogs finally avoided the electric shocks on their own. See Peterson, Maier, and Seligman, *Learned Helplessness*, 27.

103. Overmier and Seligman, "Effects of Inescapable Shock," 33.

104. Seligman, Maier, and Geer, "Alleviation of Learned Helplessness in the Dog," 258, 261.

105. See Lyn Y. Abramson, Martin E. P. Seligman, and John D. Teasdale, "Learned Helplessness in Humans: Critique and Reformulation," *Journal of Abnormal Psychology* 87 (February 1978): 51.

106. See Seligman, Maier, and Geer, "Alleviation of Learned Helplessness in the Dog," 258. Also see Bruno Bettelheim, *The Informed Heart* (New York: Free Press, 1960).

107. Martin E. P. Seligman, "Learned Helplessness," *Annual Review of Medicine* 23 (February 1972): 411

108. See Carol S. Dweck and N. Dickon Reppucci, "Learned Helplessness and Reinforcement Responsibility in Children," *Journal of Personality and Social Psychology* 25 (January 1973): 109–16. The forty elementary schoolchildren in this study were all from a "lower middle-class suburb of New Haven, Connecticut," but the analysis did not examine class. Rather, it focused on connections between learned helplessness, academic achievement, and gender (111). Gender differences and helplessness was the direction Dweck pursued in further experiments. See Carol S. Dweck and Ellen S. Bush, "Sex Differences in Learned Helplessness: I. Differential Debilitation with Peer and Adult Evaluators," *Developmental Psychology* 12 (March 1976): 147–56; and Carol S. Dweck, William Davidson, Sharon Nelson, and Bradley Enna, "Sex Differences in Learned Helplessness: II. The Contingencies of Evaluative Feedback in the Classroom and III. An Experimental Analysis," *Developmental Psychology* 14 (May 1978): 268–76.

109. Martin E. P. Seligman, *Helplessness: On Depression, Development, and Death* (San Francisco: W. H. Freeman, 1975), 45, 135, 155.

110. Seligman, 154, 159–61.

111. Seligman, 83, 135–36, 163.

112. On the blurring of learned helplessness theory with locus of control research, see Herbert M. Lefcourt, *Locus of Control: Current Trends in Theory and Research* (New York: John Wiley and Sons, 1976). For an example of learned helplessness becoming a new reference for race, class, and education, see John R. Weisz, "Learned Helplessness in Black and White Children Identified by Their Schools as Retarded and Nonretarded: Performance Deterioration in Response to Failure," *Developmental Psychology* 17 (July 1981): 499–508.

Chapter 2

1. Robert Maynard, "Omaha Pupils Given 'Behavior' Drugs," *Washington Post*, June 29, 1970, A1, A8. Also see "Use of Child Behavior Drugs Studied," *New York Times*, June 30, 1970, 17.

2. "Treatment for Fidgety Kids? A Story of Drug Use in Omaha Kicks Up National Fuss with Some Racial Overtones," *National Observer*, July 6, 1970. Reprinted in *Journal of Learning Disabilities* 4 (November 1971): 67.

3. Ernest Chambers quoted in Harlan Vinnedge, "Drugs for Children," *New Republic*, March 13, 1971, 13.

4. "Pep Pills for Pupils," *Newsweek*, July 13, 1970.

5. Maynard, "Omaha Pupils Given 'Behavior' Drugs," A8.

6. Howard L. Millman, "Minimal Brain Dysfunction in Children — Evaluation and Treatment," *Journal of Learning Disabilities* 3 (February 1970): 35.

7. Dorothea M. Ross and Sheila A. Ross, *Hyperactivity: Research, Theory, and Action* (New York: John Wiley and Sons, 1976), 16.

8. Mark A. Stewart, "Hyperactive Children," *Scientific American*, April 1970, 94.

9. See, for instance, Nicolas Rasmussen, *On Speed: The Many Lives of Amphetamine* (New York: New York University Press, 2008), 216; Matthew Smith, *Hyperactive: The Controversial History of ADHD* (London: Reaktion Books, 2012), 119; and Russell A. Barkley, "History of ADHD," in *Attention-Deficit Hyperactivity Disorder: A Handbook for Diagnosis and Treatment*, ed. Russell A. Barkley (New York: Guilford Press, 2015), 14.

10. The reported figure of 5 to 10 percent "referred to the proportion of children believed to be in need of medical help for behavioral problems," not to children on stimulant drugs. Amelia Buttress, "The Chemical Lives of 'Children'" (Ph.D. diss., Johns Hopkins University, 2014), 201.

11. Rick Mayes and Adam Rafalovich have written that "parents were not being coerced into accepting drug therapy" in the Omaha experiment of 1970. See Rick Mayes and Adam Rafalovich, "Suffer the Restless Children: The Evolution of ADHD and Paediatric Stimulant Use, 1900–80," *History of Psychiatry* 18 (December 2007): 450. This conclusion is disputed by anecdotal evidence from contemporaneous observers. Journalist and activist Nat Hentoff, for instance, reported on the case of Mr. and Mrs. Howard Curtis, "two black parents in Omaha," who were the recipients of "persistent pressure" from their ten-year-old son's teacher to have him "examined with the aim of putting him on behavior modification drugs." Nat Hentoff, "The Drugged Classroom," *Evergreen Review* 85 (December 1970): 32.

12. Edward J. Comstock, "The End of Drugging Children: Toward the Genealogy of the ADHD Subject," *Journal of the History of the Behavioral Sciences* 47 (Winter 2011): 49. Comstock is quoting from Helen Schneider and Daniel Eisenberg, "Who Receives a Diagnosis of Attention-Deficit/ Hyperactivity Disorder in the United States Elementary School Population?," *Pediatrics* 117 (April 2006): 601.

13. A good example is the excellent research by Ilina Singh into how the marketing strategies for Ritalin pursued aggressively in the early 1970s targeted white and middle-class mothers to treat their sons. See Ilina Singh, "Not Just Naughty: 50 Years of Stimulant Drug Advertising," in *Medicating Modern America: Prescription Drugs in History*, ed. Andrea Tone and Elizabeth Siegel Watkins (New York: New York University Press, 2007), 145.

14. Leon Eisenberg, "The Sins of the Fathers: Urban Decay and Social Pathology," *American Journal of Orthopsychiatry* 32 (January 1962): 5, 9.

15. Leon Eisenberg, Roy Lachman, Peter A. Molling, Arthur Lockner, James D. Mizelle, and C. Keith Conners, "A Psychopharmacologic Experiment in a Training School for Delinquent Boys: Methods, Problems, Findings," *American Journal of Orthopsychiatry* 33 (April 1963): 445.

16. C. Keith Conners and Leon Eisenberg, "The Effects of Methylphenidate on Symptomatology and Learning in Disturbed Children," *American Journal of Psychiatry* 120 (November 1963): 460.

17. Rick Mayes, Catherine Bagwell, and Jennifer Erkulwater, *Medicating Children: ADHD and Pediatric Mental Health* (Cambridge, Mass.: Harvard University Press, 2009), 60.

18. C. Keith Conners, Leon Eisenberg, and Lawrence Sharpe, "Effects of Methylphenidate (Ritalin) on Paired-Associate Learning and Porteus Maze Performance in Emotionally Disturbed Children," *Journal of Consulting Psychology* 28 (February 1964): 15.

19. Conners, Eisenberg, and Sharpe, 20–21.

20. C. Keith Conners, Leon Eisenberg, and Avner Barcai, "Effect of Dextroamphetamine on

Children: Studies on Subjects with Learning Disabilities and School Behavior Problems," *Archives of General Psychiatry* 17 (October 1967): 484.

21. Leon Eisenberg and C. Keith Conners, *The Effect of Headstart on Developmental Processes* (Washington, D.C.: U.S. Department of Health, Education & Welfare, Office of Economic Opportunity, 1966), 6. Also see David A. Waller and C. Keith Conners, *A Follow-Up Study of Intelligence Changes in Children Who Participated in Project Headstart* (Washington, D.C.: U.S. Department of Health, Education and Welfare, Office of Economic Opportunity, 1966).

22. It was surely no coincidence that Eisenberg would later be the key figure in inaugurating and sustaining affirmative action at Harvard Medical School from 1968 on.

23. C. Keith Conners, "Symptom Patterns in Hyperkinetic, Neurotic, and Normal Children," *Child Development* 41 (September 1970): 678, 680–81.

24. C. Keith Conners, "Psychological Effects of Stimulant Drugs in Children with Minimal Brain Dysfunction," *Pediatrics* 49 (May 1972): 708.

25. Leon Eisenberg, "The Clinical Use of Stimulant Drugs in Children," *Pediatrics* 49 (May 1972): 712.

26. Leon Eisenberg, "Principles of Drug Therapy in Child Psychiatry with Special Reference to Stimulant Drugs," *American Journal of Orthopsychiatry* 41 (April 1971): 375.

27. Charles Bradley, "Benzedrine and Dexedrine in the Treatment of Children's Behavior Disorders," *Pediatrics* 5 (January 1950): 24–25. Here, Bradley was quoting from his own prior research findings. See Charles Bradley and Margaret Bowen, "Amphetamine (Benzedrine) Therapy of Children's Behavior Disorders," *American Journal of Orthopsychiatry* 11 (January 1941): 95.

28. Charles Bradley, "The Behavior of Children Receiving Benzedrine," *American Journal of Psychiatry* 94 (November 1937): 582.

29. Maurice W. Laufer and Eric Denhoff, "Hyperkinetic Behavior Syndrome in Children," *Journal of Pediatrics* 50 (April 1957): 463.

30. Matthew Smith, "Putting Hyperactivity in Its Place: Cold War Politics, the Brain Race and the Origins of Hyperactivity in the United States, 1957–68," in *Locating Health: Historical and Anthropological Investigations of Health and Place*, ed. Erika Dyck and Christopher Fletcher (London: Pickering and Chatto, 2011), 60.

31. "The Hyperactive Child," *Juvenile Court Judges Journal* 12 (January 1962): 14.

32. Mark A. Stewart, Ferris N. Pitts, Alan G. Craig, and William Dieruf, "The Hyperactive Child Syndrome," *American Journal of Orthopsychiatry* 36 (October 1966): 864, 866. A mother testified to the U.S. Congress in 1970, for instance, that her son Tommy "was not only bright, but almost brilliant," even while he also often underwent a "Jekyll-Hyde transformation"— until he was placed on Ritalin. See the letter of Mrs. Frederick N. Kelly reprinted in *Federal Involvement in the Use of Behavior Modification Drugs on Grammar School Children of the Right to Privacy Inquiry: Hearings before a Subcommittee of the Committee on Government Operations, House of Representatives*, 91st Congress, 2nd Session, September 29, 1970 (Washington, D.C.: U.S. Government Printing Office, 1970), 120–21.

33. See, for instance, Gabrielle Weiss, John Werry, Klaus Minde, Virginia Douglas, and Donald Sykes, "Studies on the Hyperactive Child — V. The Effects of Dextroamphetamine and Chlorpromazine on Behaviour and Intellectual Functioning," *Journal of Child Psychology and Psychiatry* 9 (December 1968): 145.

34. Sam D. Clements, *Minimal Brain Dysfunction in Children*, National Institute of Neurological Diseases and Blindness, Monograph No. 3 (Washington, D.C.: U.S. Department of Health, Education, and Welfare, 1966), 6–10.

35. Clements, 9.

36. Mical Raz, *What's Wrong with the Poor? Psychiatry, Race, and the War on Poverty* (Chapel Hill: University of North Carolina Press, 2013), 122. Also see Maris A. Vinovskis, *The Birth of*

Head Start: Preschool Education Policies in the Kennedy and Johnson Administrations (Chicago: University of Chicago Press, 2005).

37. See Maurice W. Laufer, Eric Denhoff, and Gerald Solomons, "Hyperkinetic Impulse Disorder in Children's Behavior Problems," *Psychosomatic Medicine* 19 (January 1957): 38–49; Laufer and Denhoff, "Hyperkinetic Behavior Syndrome in Children"; and the extensive list of terms mentioned in Clements, *Minimal Brain Dysfunction in Children*, 9.

38. Stewart, Pitts, Craig, and Dieruf, "Hyperactive Child Syndrome," 863.

39. Quoted in Roger Signor, "Hyperactive Children," *Washington University Magazine*, Winter 1967, 23.

40. Similarly, an epidemiological study published in 1967 that surveyed over three hundred second-graders living in rural Vermont concluded that "10% of the children seemed to meet the definition of hyperkinesis." While no mention was made of the racial (or class) background of these rural Vermont children, the study did observe that most of the hyperkinetic children in the study had "improved greatly on amphetamines" and that those "who did not respond to amphetamines . . . did respond positively to Ritalin (methyl-phenidate)." Hans R. Huessey, "Study of the Prevalence and Therapy of the Hyperkinetic Syndrome in Public School Children in Rural Vermont," *Acta Paedopsychiatrica* 34 (1967): 131–32.

41. John M. Krager and Daniel J. Safer, "Type and Prevalence of Medication Used in the Treatment of Hyperactive Children," *New England Journal of Medicine* 291 (November 1974): 1120.

42. Urie Bronfenbrenner, "The Psychological Costs of Quality and Equality in Education," *Child Development* 38 (December 1967): 913. Bronfenbrenner went on to argue that "integration cannot repair a damaged brain, supply a father, equip a home with books, or alter a family's values, speech habits, and patterns of child rearing. Thus, in many cases, the Negro child in the integrated classroom is, and continues to be, intellectually retarded, unable to concentrate, unmotivated to learn; at first apathetic, but as he gets older, becoming resentful, rebellious, and delinquency-prone" (918).

43. Murray M. Kappelman, Eugene Kaplan, and Robert L. Ganter, "A Study of Learning Disorders among Disadvantaged Children," *Journal of Learning Disabilities* 2 (May 1969): 27, 31–33.

44. Edith H. Grotberg, "Neurological Aspects of Learning Disabilities: A Case for the Disadvantaged," *Journal of Learning Disabilities* 3 (June 1970): 30–31.

45. Daniel P. Hallahan, "Cognitive Styles — Preschool Implications for the Disadvantaged," *Journal of Learning Disabilities* 3 (January 1970): 9. Hallahan soon thereafter prepared a leading text on learning disabilities. See Daniel P. Hallahan and James M. Kauffman, *Introduction to Learning Disabilities: A Psycho-Behavioral Approach* (Englewood Cliffs, N.J.: Prentice-Hall, 1976).

46. Lester Tarnopol, "Delinquency and Minimal Brain Dysfunction," *Journal of Learning Disabilities* 3 (April 1970): 27.

47. Hallahan, "Cognitive Styles," 7.

48. Dominic Amante, Phillip H. Margules, Donna M. Hartman, Delores B. Storey, and Lewis John Weeber, "The Epidemiological Distribution of CNS Dysfunction," *Journal of Social Issues* 26 (Autumn 1970): 106, 121, 128.

49. Arthur R. Jensen, *Educability and Group Differences* (New York: Harper and Row, 1973), 343, 348–49.

50. John L. Horn, review of *Educability and Group Differences*, by Arthur R. Jensen, *American Journal of Psychology* 87 (September 1974): 546.

51. Quoted in Robert Reinhold, "Rx for Child's Learning Malady," *New York Times*, July 3, 1970, 28.

52. Maurice W. Laufer, "Medications, Learning, and Behavior," *Phi Delta Kappan* 52 (November 1970): 169–70; Paul H. Wender, "The Minimal Brain Dysfunction Syndrome," *Annual Review of Medicine* 26 (1975): 57.

53. Nicholas von Hoffman, "Student Pep Talk," *Washington Post*, July 22, 1970.

54. Lester Grinspoon and Susan B. Singer, "Amphetamines in the Treatment of Hyperkinetic Children," *Harvard Educational Review* 43 (November 1973): 547.

55. Peter Conrad, *Identifying Hyperactive Children: The Medicalization of Deviant Behavior* (Lexington, Mass.: Lexington Books, 1976), 4. Also see Peter Conrad, "The Discovery of Hyperkinesis: Notes on the Medicalization of Deviant Behavior," *Social Problems* 23 (October 1975): 12–21.

56. T. Berry Brazelton, "The Children Who Can't Sit Still," *Redbook*, August 1972, 70–71, 183, 185; Bruno Bettelheim, "Bringing Up Children," *Ladies Home Journal*, February 1973, 28, 30, 130–31.

57. Jane E. Brody, "When a Child Has Trouble Learning," *Woman's Day*, January 1973, 18, 114, 116, 118, 120; Kenneth L. Woodward, "When Your Child Can't Read," *McCall's*, February 1973, 48, 50, 52, 57; William Cole, "Breaking the Hidden Barrier to Learning," *Parents Magazine*, September 1972, 55, 98–100; Bert Kruger Smith, "Free at Last to Learn," *Reader's Digest*, June 1972, 41–48; Dr. Herbert Neff, "The Child Who Wasn't Retarded," *PTA Magazine*, May 1973, 21–23.

58. Mark A. Stewart and Sally Wendkos Olds, *Raising a Hyperactive Child* (New York: Harper and Row, 1973), 238. Also see Paul H. Wender, *The Hyperactive Child: A Handbook for Parents* (New York: Crown, 1973).

59. Susan Hunsinger, "School Storm: Drugs for Children," *Christian Science Monitor*, October 31, 1970. The text of the Students for a Democratic Society flyer can be found in *Journal of Learning Disabilities* 4 (November 1971): 7.

60. Charles Witter, "Drugging and Schooling," in *Children and Their Caretakers*, ed. Norman K. Denzin (New Brunswick, N.J.: Transaction Books, 1973), 71. Reprinted from *Trans-action* (July/August 1971).

61. Alan F. Charles, "Drugs for Hyperactive Children: The Case of Ritalin," *New Republic*, October 23, 1971, 18.

62. Nat Hentoff, "Order in the Classroom!," *Village Voice*, December 3, 1970.

63. Quoted in Nat Hentoff, "The Educational-Psychiatric-Psychological Complex," *Village Voice*, June 22, 1972, 14.

64. Richard A. Johnson, James B. Kenney, and John B. Davis, "Developing School Policy for Use of Stimulant Drugs for Hyperactive Children," *School Review* 85 (November 1976): 91.

65. William Banks, "Drugs, Hyperactivity, and Black Schoolchildren," *Journal of Negro Education* 45 (Spring 1976): 156–57.

66. "An Interview with Marian Wright Edelman," *Harvard Educational Review* 44 (February 1974): 64.

67. Melvin H. King quoted in "MBD, Drug Research and the Schools: A Conference on Medical Responsibility and Community Control," *Hastings Center Report* 6 (June 1976): 20.

68. A. Wade Boykin, "Psychological/Behavioral Verve in Academic/Task Performance: Pretheoretical Considerations," *Journal of Negro Education* 47 (Autumn 1978): 353. Here Boykin drew on psychologist Marcia Guttentag's observations that black children moved significantly more often than did white children. Marcia Guttentag, "Negro-White Differences in Children's Movement," *Perceptual and Motor Skills* 35 (October 1972): 435–36.

69. Jane R. Mercer, "Socio-cultural Factors in Labeling Mental Retardates," *Peabody Journal of Education* 48 (April 1971): 189. Also see Jane R. Mercer, *Labeling the Mentally Retarded: A Clinical and Social System Perspective on Mental Retardation* (Berkeley: University of California Press, 1973).

70. Jean Kealy and John McLeod, "Learning Disability and Socioeconomic Status," *Journal of Learning Disabilities* 9 (November 1976): 65–66.

71. Peter Schrag and Diane Divoky, *The Myth of the Hyperactive Child and Other Means of Child Control* (New York: Pantheon Books, 1975), 55. Also see Peter Schrag, "Readin', Writin' (and Druggin')," *New York Times*, October 19, 1975.

72. J. David Smith and Edward A. Polloway, "Learning Disabilities: Individual Needs or Categorical Concerns?," *Journal of Learning Disabilities* 12 (October 1979): 28.

73. Kenneth A. Kavale, "Learning Disability and Cultural-Economic Disadvantage: The Case for a Relationship," *Learning Disability Quarterly* 3 (Summer 1980): 98. Also see E. Hall, "Special Miseducation: The Politics of Special Education," *Inequality in Education* 4 (1970): 17–29.

74. Testimony of Dr. Ronald Lipman, Chief, Clinical Studies Section, FDA, *Federal Involvement in the Use of Behavior Modification Drugs on Grammar School Children of the Right to Privacy Inquiry*, 16.

75. Cited in Robert L. Sprague and Kenneth D. Gadow, "The Role of the Teacher in Drug Treatment," *School Review* 85 (November 1976): 111.

76. Krager and Safer, "Type and Prevalence of Medication Used in the Treatment of Hyperactive Children," 1119.

77. "Classroom Pushers," *Time*, February 26, 1973. In 1974, a report in the *New England Journal of Medicine* stated that "there appear to be at least 300,000 children in elementary schools in the United States on psychotropic medication for school difficulties." Krager and Safer, "Type and Prevalence of Medication Used in the Treatment of Hyperactive Children," 1120.

78. Schrag and Divoky, *Myth of the Hyperactive Child*, 249–50, xii. These figures have been challenged as "highly speculative." See Sprague and Gadow, "Role of the Teacher in Drug Treatment," 113.

79. Jon Tobiessen and Harris E. Karowe, "A Role for the School in the Pharmacological Treatment of Hyperkinetic Children," *Psychology in the Schools* 6 (October 1969): 340.

80. Hunsinger, "School Storm."

81. Wender, "Minimal Brain Dysfunction Syndrome," 52.

82. Lester Grinspoon and Peter Hedblom, *The Speed Culture: Amphetamine Use and Abuse in America* (Cambridge, Mass.: Harvard University Press, 1975), 230.

83. Grinspoon and Singer, "Amphetamines in the Treatment of Hyperkinetic Children," 546.

84. Ben F. Feingold, *Why Your Child Is Hyperactive* (New York: Random House, 1974), 66–67. See also Ben F. Feingold, "Hyperkinesis and Learning Disabilities Linked to the Ingestion of Artificial Food Colors and Flavors," *Journal of Learning Disabilities* 9 (November 1976): 19–27. For a contemporaneous and critical assessment of the Feingold diet, see C. Keith Conners, *Food Additives and Hyperactive Children* (New York: Plenum, 1980). For a historical overview, see Matthew Smith, "Into the Mouths of Babes: Hyperactivity, Food Additives, and the Reception of the Feingold Diet," in *Health and the Modern Home*, ed. Mark Jackson (New York: Routledge, 2007), 304–21.

85. Virginia I. Douglas, "Stop, Look and Listen: The Problem of Sustained Attention and Impulse Control in Hyperactive and Normal Children," *Canadian Journal of the Behavioral Sciences* 4 (1972): 259–82; Donald H. Sykes, Virginia I. Douglas, and Gert Morgenstern, "The Effect of Methylphenidate (Ritalin) on Sustained Attention in Hyperactive Children," *Psychopharmacologia* 25 (1972): 262–74.

86. Although an ADHD diagnosis in *DSM-5* does require "pervasiveness across settings," "impairment in one or more major domains of life functioning," and "early onset of symptoms (before age 13)." See Walter Roberts, Richard Milich, and Russell A. Barkley, "Primary Symptoms, Diagnostic Criteria, Subtyping, and Prevalence of ADHD," in Barkley, *Attention-Deficit Hyperactivity Disorder*, 71.

87. James M. Swanson, Marc Lerner, and Lillie Williams, "More Frequent Diagnosis of Attention Deficit-Hyperactivity Disorder" (letter), *New England Journal of Medicine* 333 (October 1995): 944. Also see Claudia Wallis, "Life in Overdrive: An Epidemic of Attention Deficit Disorder," *Time*, July 18, 1994, 42–50.

88. Barbara Bloom, Robin A. Cohen, and Gulnur Freeman, *Summary Health Statistics for U.S. Children: National Health Interview Survey, 2009* (Washington, D.C.: National Center for Health Statistics, 2010), 5.

89. These debates were especially divisive in the late 1990s. See Sydney Walker, *The Hyperactivity Hoax: How to Stop Drugging Your Child and Find Real Medical Help* (New York: St. Martin's Press, 1998). Here Walker wrote, "Hyperactivity is not a disease. It's a hoax perpetrated by doctors

who have no idea what's really wrong with these children" (5). Also see Malcolm Gladwell, "Running From Ritalin," *New Yorker*, February 15, 1999, 80–84. Here Gladwell concluded, "Modernity didn't create A.D.H.D. It revealed it" (84).

90. Klaus W. Lange, Susanne Reichl, Katharina M. Lange, Lara Tucha, and Oliver Tucha, "The History of Attention Deficit Hyperactivity Disorder," *ADHD Attention Deficit and Hyperactivity Disorders* 2 (December 2010): 253.

91. Joel T. Nigg, *What Causes ADHD? Understanding What Goes Wrong and Why* (New York: Guilford Press, 2006), 17.

92. Lawrence H. Diller, "The Run on Ritalin: Attention Deficit Disorder and Stimulant Treatment in the 1990s," *Hastings Center Report* 26 (March–April 1996): 13.

93. Quoted in Alan Schwarz, "The Selling of Attention Deficit Disorder," *New York Times*, December 14, 2013.

94. Quoted in Craig S. Lerner, "'Accommodations' for the Learning Disabled: A Level Playing Field or Affirmative Action for Elites?," *Vanderbilt Law Review* 57 (2004): 1069.

95. Dyan Machan, "An Agreeable Affliction," *Forbes*, August 12, 1996, 151.

96. Ruth Shalit, "Defining Disabilities Down," *New Republic*, August 25, 1997, 22. For Vint Lawrence's cartoon, see 19.

97. Mary Eberstadt, "Why Ritalin Rules," *Policy Review* 94 (April–May 1999): 34.

98. Valerie J. Samuel, Shannon Curtis, Ayanna Thornell, Patricia George, Andrea Taylor, Deborah Ridley Brome, Joseph Biederman, and Stephen V. Faraone, "The Unexplored Void of ADHD and African-American Research: A Review of the Literature," *Journal of Attention Disorders* 1 (January 1997): 205.

99. Torri W. Miller, Joel T. Nigg, and Robin L. Miller, "Attention Deficit Hyperactivity Disorder in African American Children: What Can Be Concluded from the Past Ten Years?," *Clinical Psychology Review* 29 (February 2009): 84. Also see Judy C. Davison and Donna Y. Ford, "Perceptions of Attention Deficit Hyperactivity Disorder in One African American Community," *Journal of Negro Education* 70 (Fall 2001): 264–74; and Regina Bussing, Faye A. Gary, Terry L. Mills, and Cynthia Wilson Garvan, "Parental Explanatory Models of ADHD: Gender and Cultural Variations," *Social Psychiatry and Psychiatric Epidemiology* 38 (October 2003): 563–75.

Chapter 3

1. Neuropsychologist Marcel Kinsbourne is credited with coining the term "dichotomania" to denote split-brain theorizing "in its excesses." See David Galin, "The Two Modes of Consciousness and the Two Halves of the Brain," in *Consciousness, Brain, States of Awareness, and Mysticism*, ed. David Goleman and Richard J. Davidson (New York: Harper and Row, 1979), 23.

2. Henry Mintzberg, "Planning on the Left Side and Managing on the Right," *Harvard Business Review*, July–August 1976, 57–58.

3. Julian Jaynes, *The Origins of Consciousness in the Breakdown of the Bicameral Mind* (Boston: Houghton Mifflin, 1976), 107, 404–5.

4. The book was not always a critical success. See, for example, the review that labeled Jaynes's hypothesis a "literary fantasy with a dash of apparent evidence from neurobiology, achieving originality at the cost of common sense." Bernard D. Davis, "Speculating on the Brain," *Hastings Center Report* 8 (April 1978): 35.

5. Carl Sagan, *The Dragons of Eden: Speculations on the Evolution of Human Intelligence* (New York: Ballantine Books, 1977), 177.

6. Betty Edwards, *Drawing on the Right Side of the Brain: A Course in Enhancing Creativity and Artistic Confidence* (Los Angeles: J. P. Tarcher, 1979), 37.

7. Arthur R. Jensen, "Patterns of Mental Ability and Socioeconomic Status," *Proceedings of the National Academy of Sciences of the United States of America* 60 (August 15, 1968): 1330.

8. In the preface to *Genetics and Education*, Jensen describes for close to seventy pages the "storm of ideologically, often politically, motivated protests, misinterpretations, and vilifications" his writing on race and intelligence had prompted. See Arthur R. Jensen, *Genetics and Education* (New York: Harper and Row, 1972), 1.

9. On the nineteenth-century history of the split-brain paradigm and Liepmann's findings, see Anne Harrington, "Nineteenth-Century Ideas on Hemisphere Differences and 'Duality of Mind,'" *Behavioral and Brain Sciences* 8 (December 1985): 617–34; and Anne Harrington, "Unfinished Business: Models of Laterality in the Nineteenth Century," in *Brain Asymmetry*, ed. Richard J. Davidson and Kenneth Hugdahl (Cambridge, Mass.: MIT Press, 1995), 3–27.

10. See William P. van Wagenen and R. Yorke Herren, "Surgical Division of the Commissural Pathways in the Corpus Callosum: Relation to Spread of an Epileptic Attack," *Archives of Neurology and Psychiatry* 44 (October 1940): 740–59; and Andrew J. Akelaitis, "Studies on the Corpus Callosum: II. The Higher Visual Functions in Each Homonymous Field Following Complete Section of the Corpus Callosum," *Archives of Neurology and Psychiatry* 45 (May 1941): 788–96.

11. See, for instance, Ronald E. Myers and R. W. Sperry, "Interhemispheric Communication through the Corpus Callosum: Mnemonic Carry-Over between the Hemispheres," *Archives of Neurology and Psychiatry* 80 (September 1958): 298–303.

12. Roger W. Sperry, "Some General Aspects of Interhemispheric Integration," in *Interhemispheric Relations and Cerebral Dominance*, ed. Vernon B. Mountcastle (Baltimore: Johns Hopkins Press, 1962), 46, 47.

13. R. W. Sperry, "Cerebral Organization and Behavior: The Split Brain Behaves in Many Respects Like Two Separate Brains, Providing New Research Possibilities," *Science*, June 2, 1961, 1749.

14. Mountcastle, preface to *Interhemispheric Relations and Cerebral Dominance*, v.

15. M. S. Gazzaniga, J. E. Bogen, and R. W. Sperry, "Some Functional Effects of Sectioning the Cerebral Commissures in Man," *Proceedings of the National Academy of Sciences of the United States of America* 48 (October 15, 1962): 1766, 1767.

16. R. W. Sperry, "The Great Cerebral Commissure," *Scientific American*, January 1964, 42, 52.

17. Michael S. Gazzaniga, "The Split Brain in Man," *Scientific American*, August 1967, 24. Also see Michael S. Gazzaniga, *The Bisected Brain* (New York: Appleton-Century-Crofts, 1970).

18. Jerre Levy-Agresti and R. W. Sperry, "Differential Perceptual Capacities in Major and Minor Hemispheres," *Proceedings of the National Academy of Sciences of the United States of America* 61 (November 15, 1968): 1151.

19. See Jerre Levy, "Lateral Specialization of the Human Brain: Behavioral Manifestation and Possible Evolutionary Basis," in *The Biology of Behavior*, ed. John A. Kiger Jr. (Corvallis: Oregon State University Press, 1972), 159–80.

20. See, for instance, Sandra F. Witelson, "Sex and the Single Hemisphere: Specialization of the Right Hemisphere for Spatial Processing," *Science*, July 30, 1976, 425–27; and Daniel Goleman, "Special Abilities of the Sexes: Do They Begin in the Brain?," *Psychology Today*, November 1978, 48–59.

21. Joseph E. Bogen, "The Corpus Callosum, the Other Side of the Brain and Pharmacologic Opportunity," in *Drugs and Cerebral Function*, ed. W. L. Smith (Springfield, Ill.: C. C. Thomas, 1970), 11.

22. Joseph E. Bogen, "The Other Side of the Brain I: Dysgraphia and Dyscopia Following Cerebral Commissurotomy," *Bulletin of the Los Angeles Neurological Societies* 34 (April 1969): 105.

23. Joseph E. Bogen, "The Other Side of the Brain II: An Appositional Mind," *Bulletin of the Los Angeles Neurological Societies* 34 (July 1969): 156. This article by Bogen went on to achieve a sort of cult status after Philip K. Dick chose to quote from it directly in his science fiction novel *A Scanner Darkly* (1977).

24. Joseph E. Bogen, "The Other Side of the Brain III: The Corpus Callosum and Creativity," *Bulletin of the Los Angeles Neurological Societies* 34 (October 1969): 200–201, 202.

25. H. W. Gordon and J. E. Bogen, "Hemispheric Lateralization of Singing after Intracarotid Sodium Amylobarbitone," *Journal of Neurology, Neurosurgery, and Psychiatry* 37 (June 1974): 733.

26. Maya Pines, "We Are Left-Brained or Right-Brained," *New York Times Magazine*, September 9, 1973, 32, 126.

27. See Stephen E. Wald, "Minds Divided: Science, Spirituality, and the Split Brain in American Thought" (Ph.D. diss., University of Wisconsin-Madison, 2008).

28. Robert E. Ornstein, *The Psychology of Consciousness* (San Francisco: W. H. Freeman, 1972), 178.

29. Robert E. Ornstein, "Two for One," *New York Times*, August 9, 1973, 35.

30. See Robert E. Ornstein, *The Nature of Human Consciousness: A Book of Readings* (San Francisco: W. H. Freeman, 1973).

31. David Galin and Robert Ornstein, "Lateral Specialization of Cognitive Mode: An EEG Study," *Psychophysiology* 9 (July 1972): 413, 418. Galin and Ornstein's concluding hope that their work could "enable the training of ordinary individuals to achieve more precise control over their brains' activities" was purely speculative, based on their idea that while some extant biofeedback efforts were directed to encouraging either "theta" or "alpha" activity, their own "development of an index of lateralized functions" might provide a basis for targeted training of hemispheres (417–18). For a typical example of the enthusiasm of the time for biofeedback's capability for "turning on the power of your mind," see Marvin Karlins and Lewis M. Andrews, *Biofeedback: Turning on the Power of Your Mind* (Philadelphia: Lippincott, 1972).

32. Marilyn Ferguson, *The Brain Revolution: The Frontiers of Mind Research* (New York: Taplinger, 1973), 77–78.

33. Thomas H. Budzynski, "Biofeedback and the Twilight States of Consciousness," in *Consciousness and Self-Regulation: Advances in Research*, ed. Gary E. Schwartz, David Shapiro, and Richard J. Davidson (New York: Plenum Press, 1976), 364, 381. Also see Bogen, "Other Side of the Brain III."

34. Gary E. Schwartz, "Biofeedback, Self-Regulation, and the Patterning of Physiological Processes," *American Scientist*, June 1975, 320. Although the findings appeared to cast into doubt, or at least complicate, the prior assumption that the left brain was rational while the right brain was intuitive, Schwartz hypothesized that the brain's hemispheres processed positive and negative emotions asymmetrically and determined that positive emotions engaged the left hemisphere and negative emotions engaged the right. See Gary E. Schwartz et al., "Lateralized Facial Muscle Response to Positive and Negative Emotional Stimuli," *Psychophysiology* 16 (November 1979): 561–71. *Psychology Today* in 1981 reported that there was "inescapable" evidence "for a hemispheric division of emotional labor" and argued that the "the right hemisphere is involved in negative feelings and their expression, while the left is associated with positive feelings and their expression." Marcel Kinsbourne, "Sad Hemisphere, Happy Hemisphere," *Psychology Today*, May 1981, 92. By 1982 Richard Davidson was pursuing influential research into facial expressions as a means of establishing correlations between differential hemispheric activation and affective states. With the assistance of developmental psychologist Nathan Fox, Davidson designed a study on the "psychophysiology of emotional development." This clever experiment involved more than three dozen infants (who sat on their mothers' laps during the tests) while Davidson and Fox attached each child to an EEG and measured his or her "response to positive and negative affective stimuli" (in this instance, film clips of an actress either laughing or sobbing). What the EEG patterns revealed — here confirming Schwartz's findings — was heightened right hemisphere activation when the affective stimulus was negative (i.e., the actress cried) and elevated left frontal activity when the emotional stimulus was positive (i.e., the actress laughed). Davidson and Fox concluded that these findings with infants supported a hypothesis that "affective lateralization" existed and that "such asymmetries" were fully developed in children already "in the first year of life." Not only were the hemispheres of the brain highly specialized in terms of how they handled emotional stimuli, but such specialization appeared to be hard-wired from birth. See Richard J. Davidson and Nathan A. Fox, "Asymmetrical Brain Activity Discriminates between Positive and Negative Affective Stimuli in Human Infants," *Science*, December 17, 1982, 1235, 1236. Davidson would later enthuse in his intellectual autobiography that it had been with the publication of the first in this series of EEG studies with infants that "the field

of affective neuroscience — the study of the brain basis of emotion — was launched." See Richard J. Davidson with Sharon Begley, *The Emotional Life of Your Brain: How Its Unique Patterns Affect the Way You Think, Feel, and Live — and How You Can Change Them* (New York: Penguin, 2012), 36. For a powerful exposé of developments in the field of affect studies, see Ruth Leys, *The Ascent of Affect: Genealogy and Critique* (Chicago: University of Chicago Press, 2017).

35. Albert Rosenfeld and Kenneth W. Klivington, "Inside the Brain: The Last Great Frontier," *Saturday Review*, August 9, 1975, 14.

36. Linda Rogers, Warren TenHouten, Charles D. Kaplan, and Martin Gardiner, "Hemispheric Specialization of Language: An EEG Study of Bilingual Hopi Indian Children," *International Journal of Neuroscience* 8 (1977): 1–6. Also see George W. Hynd, Anne Teeter, and Jennifer Stewart, "Acculturation and the Lateralization of Speech in the Bilingual Native American," *International Journal of Neuroscience* 11 (1980): 1–7.

37. Steve Scott, George W. Hynd, Lester Hunt, and Wendy Weed, "Cerebral Speech Lateralization in the Native American Navajo," *Neuropsychologia* 17 (1979): 91.

38. See George W. Hynd and Steve A. Scott, "Propositional and Appositional Modes of Thought and Differential Cerebral Speech Lateralization in Navajo Indian and Anglo Children," *Child Development* 51 (September 1980): 911.

39. Bogen, "Other Side of the Brain III," 202.

40. Warren D. TenHouten, *Cognitive Styles and the Social Order* (Washington, D.C.: Office of Economic Opportunity, July 1971), 1, 46–48, 53, 135, 137. Also see the chapter "Language in Inquiry" in Warren D. TenHouten, *Science and Its Mirror Image: A Theory of Inquiry* (New York: Harper and Row, 1973), 90–108.

41. Thom Borowy and Ronald Goebel, "Cerebral Lateralization of Speech: The Effects of Age, Sex, Race, and Socioeconomic Class," *Neuropsychologia* 14 (1976): 363, 368.

42. Arthur R. Jensen, "How Much Can We Boost IQ and Scholastic Achievement?," *Harvard Educational Review* 39 (Winter 1969): 3, 82.

43. See, for instance, Martin Deutsch, "Happenings on the Way Back to the Forum: Social Science, IQ, and Race Differences Revisited," *Harvard Educational Review* 39 (September 1969): 523–57.

44. TenHouten, *Cognitive Styles and the Social Order*, 11.

45. Joseph E. Bogen, R. DeZure, Warren D. TenHouten, and J. F. Marsh Jr., "The Other Side of the Brain IV: The A/P Ratio," *Bulletin of the Los Angeles Neurological Societies* 37 (April 1972): 49–50, 58.

46. Joseph E. Bogen, "Some Educational Aspects of Hemispheric Specialization," in *The Human Brain*, ed. Merlin C. Wittrock (Englewood Cliffs, N.J.: Prentice-Hall, 1977), 146. This essay had originally been published in the *UCLA Educator* 17 (Spring 1975). Also see Andrea Lee Thompson, Joseph E. Bogen, and John F. Marsh Jr., "Cultural Hemisphericity: Evidence from Cognitive Tests," *International Journal of Neuroscience* 9 (1979): 37–43.

47. Andrea Lee Thompson and Joseph E. Bogen, "More on the Question of Cultural Hemisphericity," *Bulletin of the Los Angeles Neurological Society* 41 (July 1976): 94.

48. Strikingly, Joseph Bogen sometimes cited Jensen to make it appear as if Jensen was very much in agreement with his own perspectives, for instance quoting Jensen's call for "greater diversity of curricula, instructional methods, and educational goals and methods." See Bogen, "Some Educational Aspects of Hemispheric Specialization," 148. Yet Jensen had also declared — near the passage Bogen quoted — that "no new instructional programme has been discovered which, when applied on a large scale, has appreciably raised the scholastic achievement of disadvantaged children in relation to the majority of the school population." See Arthur R. Jensen, "Genetics and Education: A Second Look," *New Scientist*, October 12, 1972, 98.

49. Bob Samples, *The Metaphoric Mind: A Celebration of Creative Consciousness* (Reading, Mass.: Addison-Wesley, 1976), 151–52. Also see Robert E. Samples, "Are You Teaching Only One Side of the Brain?," *Learning: The Magazine for Creative Teaching*, February 1975, 25–27.

50. Esther Cappon Gray, "Brain Hemispheres and Thinking Styles," *Clearing House* 54 (November 1980): 131.

51. Gerald Gregory Jackson, "Black Psychology: An Avenue to the Study of Afro-Americans," *Journal of Black Studies* 12 (March 1982): 245–46.

52. Maya Pines, *The Brain Changers: Scientists and the New Mind Control* (New York: Harcourt Brace Jovanovich, 1973), 151.

53. Robert E. Ornstein and David Galin, "Physiological Studies of Consciousness," in *Symposium on Consciousness*, by Philip R. Lee et al. (New York: Viking, 1976), 65.

54. Michael S. Gazzaniga, "Brain—Corpus Callosum: Review of the Split Brain," *Journal of Learning Disabilities* 8 (November 1975): 569.

55. R. W. Sperry, "Lateral Specialization of Cerebral Function in the Surgically Separated Hemispheres," in *The Psychophysiology of Thinking: Studies of Covert Processes*, ed. F. J. McGuigan and R. A. Schoonover (New York: Academic Press, 1973), 209.

56. Roger W. Sperry, "Left-Brain, Right-Brain," *Saturday Review*, August 9, 1975, 33.

57. Roger Sperry, "Some Effects of Disconnecting the Cerebral Hemispheres," *Science*, September 24, 1982, 1225. For a helpful overview of the stages in Sperry's research, the scene at Caltech, and the funding streams and professional networks that nourished his productivity—but also Sperry's conviction that his split-brain work had been "over-interpreted, especially by non-scientists"—see Antonio E. Puente, "Roger W. Sperry: From Neuro-science to Neuro-philosophy," in *Pathways to Prominence in Neuropsychology: Reflections on Twentieth-Century Pioneers*, ed. Anthony Y. Stringer, Eileen L. Cooley, and Anne-Lise Christensen (New York: Taylor and Francis Books, 2002), 74.

58. Raquel E. Gur, Ruben C. Gur, and Brachia Marshalek, "Classroom Seating and Functional Brain Asymmetry," *Journal of Educational Psychology* 67 (February 1975): 151–53.

59. Madeline Hunter, "Right-Brained Kids in Left-Brained Schools," *Today's Education* 65 (November–December 1976): 45–48.

60. E. Paul Torrance, "Hemisphericity and Creative Functioning," *Journal of Research and Development in Education* 15 (Spring 1982): 36. Also see Cecil R. Reynolds and E. Paul Torrance, "Perceived Changes in Styles of Learning and Thinking (Hemisphericity) through Direct and Indirect Training," *Journal of Creative Behavior* 12 (December 1978): 245–52.

61. Leanna Landsmann, "The Brain's Division of Labor," *New York Times*, April 30, 1978, EDUC24.

62. Quoted in Robert Ornstein, "The Split and the Whole Brain," *Human Nature* 1 (May 1978): 83. Geschwind had argued already in 1974, "In an illiterate society a lack of visual-auditory associations would not seriously inconvenience anyone except in unusual situations; literacy makes this ability highly important." Norman Geschwind, "Disconnexion Syndromes in Animals and Man," in *Norman Geschwind: Selected Papers on Language and the Brain*, ed. Robert S. Cohen and Marx W. Wartofsky (Boston: D. Reidel, 1974), 229.

63. Dirk J. Bakker, "Hemispheric Differences and Reading Strategies: Two Dyslexias?," *Bulletin of the Orton Society* 29 (1979): 91. Also see John R. Kershner, "Cerebral Dominance in Disabled Readers, Good Readers, and Gifted Children: Search for a Valid Model," *Child Development* 48 (March 1977): 61–67; and Albert J. Harris, "Lateral Dominance and Reading Disability," *Journal of Learning Disabilities* 12 (May 1979): 337–43.

64. Dorothy van den Honert, "A Neuropsychological Technique for Training Dyslexics," *Journal of Learning Disabilities* 10 (January 1977): 22, 27. Also see Max Coltheart, "Deep Dyslexia: A Right Hemisphere Hypothesis," in *Deep Dyslexia*, ed. Max Coltheart, Karalyn Patterson, and John C. Marshall (Boston: Routledge and Kegan Paul, 1980), 326–80.

65. Harold W. Gordon, "The Learning Disabled Are Cognitively Right," *Topics in Learning and Learning Disabilities* 3 (April 1983): 34. Gordon wrote, "Based on functional observations, the learning disabled are locked into systems that largely subserve functions attributed to the right hemisphere. For the learning disabled, these systems are active in every cognitive task performed by the subject. For the normal learner, different systems are active for different tasks" (36–37).

66. Alan S. Kaufman, "Cerebral Specialization and Intelligence Testing," *Journal of Research and Development in Education* 12 (Winter 1979): 102–4.

67. Alan S. Kaufman, *Intelligent Testing with the WISC-R* (New York: John Wiley and Sons, 1979), 6–7, 160–61. On the vulnerability of the left hemisphere, also see David Galin, "Educating Both Halves of the Brain," *Childhood Education* 53 (October 1976): 17–20. Galin noted as well that "the organization of the brain is very plastic in children" and that therefore "the lateralization of cognitive function is still in flux after the acquisition of speech and even after the acquisition of written language." Thus "we may consider how the learning process of the two hemispheres may be affected differently by various daily schedules and school calendars, by the arrangement of the school building and the seating arrangement or lack of arrangement in the classroom, and by opportunities or restrictions on gross physical movement" (20).

68. Alan S. Kaufman, "Issues in Psychological Assessment: Interpreting the WISC-R Intelligently," in *Advances in Clinical Child Psychology*, vol. 3, ed. Benjamin B. Lahey and Alan E. Kazdin (New York: Plenum Press, 1980), 207–9.

69. Alan S. Kaufman and Nadeen L. Kaufman, *Kaufman Assessment Battery for Children: Interpretive Manual* (Circle Pines, Minn.: American Guidance Service, 1983).

70. Elizabeth O. Lichtenberger and Alan S. Kaufman, "Kaufman Assessment Battery for Children — Second Edition (KABC-II)," in *Encyclopedia of Cross-Cultural School Psychology*, ed. Caroline S. Clauss-Ehlers (New York: Springer, 2010), 557.

71. Arthur R. Jensen, "The Black-White Difference on the K-ABC: Implications for Future Tests," *Journal of Special Education* 18 (Fall 1984): 377.

72. Richard J. Herrnstein and Charles Murray, *The Bell Curve: Intelligence and Class Structure in American Life* (New York: Free Press, 1994), 290.

73. Elizabeth O. Lichtenberger and Alan S. Kaufman, "Kaufman Assessment Battery for Children (K-ABC): Recent Research," in *Advances in Cross-Cultural Assessment*, ed. Ronald J. Samuda et al. (Thousand Oaks, Calif.: SAGE, 1998), 65.

74. Alan S. Kaufman, Elizabeth O. Lichtenberger, Elaine Fletcher-Janzen, and Nadeen L. Kaufman, *Essentials of KABC-II Assessment* (Hoboken, N.J.: John Wiley and Sons, 2005), 2, 4, 20.

75. Robert D. Nebes, "Man's So-Called Minor Hemisphere," in Wittrock, *Human Brain*, 104–5. This essay was originally published in the *UCLA Educator* 17 (Spring 1975).

76. Daniel Goleman, "Split-Brain Psychology: Fad of the Year," *Psychology Today*, October 1977, 151.

77. Howard Gardner, "What We Know (and Don't Know) about the Two Halves of the Brain," *Journal of Aesthetic Education* 12 (January 1978): 113.

78. Curtis Hardyck and Randy Haapanen, "Educating Both Halves of the Brain: Educational Breakthrough or Neuromythology?," *Journal of School Psychology* 17 (Fall 1979): 219.

79. Judith Hooper and Dick Teresi, *The Three-Pound Universe* (New York: Macmillan, 1986), 223.

80. Michael C. Corballis, "Laterality and Myth," *American Psychologist* 35 (March 1980): 287.

81. Joseph E. LeDoux, "Cerebral Asymmetry and the Integrated Function of the Brain," in *Functions of the Right Cerebral Hemisphere*, ed. Andrew W. Young (New York: Academic Press, 1983), 212. Also see Michael S. Gazzaniga and Joseph E. LeDoux, *The Integrated Mind* (New York: Plenum Press, 1978).

82. Quoted in Kevin McKean, "Of Two Minds: Selling the Right Brain," *Discover*, April 1985, 34.

83. Luigi Pizzamiglio, Corradino De Pascalis, and Andrea Vignati, "Stability of Dichotic Listening Test," *Cortex* 10 (June 1974): 203–5.

84. Nancy C. Andreasen, James W. Dennert, Scott A. Olsen, and Antonio R. Damasio, "Hemispheric Asymmetries and Schizophrenia," *American Journal of Psychiatry* 139 (April 1982): 430. Also see Jaynes, *Origins of Consciousness in the Breakdown of the Bicameral Mind*, 430.

85. Walter F. McKeever, "Evidence against the Hypothesis of Right Hemisphere Language Dominance in the Native American Navajo," *Neuropsychologia* 19 (1981): 597.

86. Donna R. Vocate, "Differential Cerebral Speech Lateralization in Crow Indian and Anglo Children," *Neuropsychologia* 22 (1984): 487–94.

87. Roland D. Chrisjohn and Michael Peters, "The Pernicious Myth of the Right-Brained Indian," *Canadian Journal of Indian Education* 13 (1986): 62–71.

88. Anne Fausto-Sterling, *Myths of Gender: Biological Theories about Women and Men* (New York: Basic Books, 1985), 50. And yet these claims do persist. See Isabelle Ecuyer-Dab and Michèle Robert, "Have Sex Differences in Spatial Ability Evolved from Male Competition for Mating and Female Concern for Survival?," *Cognition* 91 (April 2004): 221–57. Also see pediatrician Daniel J. Siegel, who wrote in 2012 that female brains might be "less lateralized" than male brains due to the fact that "the role of child rearing would have necessitated more bilateralization of such functions as the use of language for communication of internal states, as well as social sensitivity and facility with nonverbal modes of communication"—while men evolved the spatial skills "necessary for hunting." Daniel J. Siegel, *The Developing Mind: How Relationships and the Brain Interact to Shape Who We Are* (New York: Guilford Press, 2012), 251.

89. See Stephen D. Krashen, *Second Language Acquisition and Second Language Learning* (New York: Pergamon Press, 1981).

90. See Renate Caine and Geoffrey Caine, *Making Connections: Teaching and the Human Brain* (Alexandria, Va.: Association for Supervision and Curriculum Development, 1991).

91. Richard Sinatra and Josephine Stahl-Gemake, *Using the Right Brain in the Language Arts* (Springfield, Ill.: Charles C. Thomas, 1983).

92. See Charmaine Della Neve, "Brain-Compatible Learning Succeeds," *Educational Leadership* 43 (October 1985): 83–85; Doris McEwen Walker, "Connecting Right and Left Brain: Increasing Academic Performance of African American Students through the Arts" (paper presented at the Annual Meeting of the National Alliance of Black School Educators, Dallas, 1995); and Renate Nummela Caine and Geoffrey Caine, "Reinventing Schools through Brain-Based Learning," *Educational Leadership* 52 (April 1995): 43–47.

93. Frances H. Rauscher, Gordon L. Shaw, and Katherine N. Ky, "Music and Spatial Task Performance," *Nature* 365 (October 1993): 611.

94. Frances H. Rauscher, Gordon L. Shaw, Linda J. Levine, Katherine N. Ky, and Eric L. Wright, "Music and Spatial Task Performance: A Causal Relationship" (paper presented at the American Psychological Association 102nd Annual Convention, Los Angeles, August 12–16, 1994), 21. Available at http://files.eric.ed.gov/fulltext/ED390733.pdf (accessed March 13, 2018).

95. Kevin Sack, "Georgia's Governor Seeks Musical Start for Babies," *New York Times*, January 15, 1998, A12.

96. Walker, "Connecting Right and Left Brain."

97. Eric Jensen, *Brain-Based Learning: The New Science of Teaching and Training* (Thousand Oaks, Calif.: Corwin Press, 1995), 6.

98. See David A. Sousa, *How the Brain Learns to Read* (Thousand Oaks, Calif.: Corwin Press, 2005); and Patricia Wolfe, *Brain Matters: Translating Research into Classroom Practice* (Alexandria, Va.: Association for Supervision and Curriculum Development, 2009).

99. Paul A. Howard-Jones, "Neuroscience and Education: Myths and Messages," *Nature Reviews Neuroscience* 15 (October 2014): 817–24.

100. Abigail L. Larrison, "Mind, Brain and Education as a Framework for Curricular Reform" (Ph.D. diss., University of California, San Diego, 2013), 65. Also see Dima Amso and B. J. Casey, "Beyond What Develops When: Neuroimaging May Inform How Cognition Changes with Development," *Current Directions in Psychological Science* 15 (February 2006): 24–29.

101. David Kaiser and W. Patrick McCray, eds., *Groovy Science: Knowledge, Innovation, and American Counterculture* (Chicago: University of Chicago Press, 2016).

102. Anne Harrington, *The Cure Within: A History of Mind-Body Medicine* (New York: W. W. Norton, 2008), 206.

103. Alan Watts, *In My Own Way: An Autobiography, 1915–1965* (New York: Pantheon Books, 1972), 248.

104. Alan Watts, *The Book: On the Taboo against Knowing Who You Are* (New York: Pantheon Books, 1966), x.

105. Leela C. Zion and Betty Lou Raker, *The Physical Side of Thinking* (Springfield, Ill.: Charles C. Thomas, 1986), 37.

Chapter 4

1. See, for instance, Howard Gardner, *Frames of Mind: The Theory of Multiple Intelligences* (New York: Basic Books, 1983); and Peter Salovey and John D. Mayer, "Emotional Intelligence," *Imagination, Cognition, and Personality* 9 (1990): 185–211.

2. Salovey and Mayer, "Emotional Intelligence," 185.

3. Daniel Goleman, *Emotional Intelligence* (New York: Bantam, 1995), 56, 81, 285, 82.

4. See Walter Mischel and Carol Gilligan, "Delay of Gratification, Motivation for the Prohibited Gratification, and Responses to Temptation," *Journal of Abnormal and Social Psychology* 69 (October 1964): 415.

5. For the first published account of the test specifically using marshmallows, see Walter Mischel, Ebbe B. Ebbesen, and Antonette Raskoff Zeiss, "Cognitive and Attentional Mechanisms in Delay of Gratification," *Journal of Personality and Social Psychology* 21 (February 1972): 204–18.

6. Among the details that varied in related experiments conducted by Mischel over the course of the 1960s, for example, were not only the kinds of rewards offered but also such matters as whether the child experienced recurrent success in the task assigned to him or her whose completion would lead to reward; whether an adult playing an interactive game with a child modeled abiding by the same rigorous standards for receiving his or her own rewards as was expected of the child; and whether a higher or lower value was placed by the child on a reward if the child could expect ultimately to receive the reward (rather than if the reward was unattainable). In the marshmallow (or penny or pretzel) experiments themselves, there were over time, and especially into the 1970s, variations in the types of supplementary stimuli offered (e.g., a slide show with images either relevant or irrelevant to the preferred reward chosen by the child) or (as a child's ability to self-distract either by fantasy and imagination or by focusing on something other than the anticipated reward came to be seen as a significant factor in the success in delaying gratification) variations in the kinds of distractions presented. Yet another variant on the marshmallow experiment involved preparatory conversations with the children about whether (and if so how) they might succeed in distracting themselves.

7. Walter Mischel, "Preference for Delayed Reinforcement: An Experimental Study of a Cultural Observation," *Journal of Abnormal and Social Psychology* 56 (January 1958): 57–61.

8. Walter Mischel, "Preference for Delayed Reinforcement and Social Responsibility," *Journal of Abnormal and Social Psychology* 62 (January 1961): 7.

9. Walter Mischel, "Delay of Gratification, Need for Achievement, and Acquiescence in Another Culture," *Journal of Abnormal and Social Psychology* 62 (May 1961): 543–52.

10. Walter Mischel, "Father-Absence and Delay of Gratification," *Journal of Abnormal and Social Psychology* 63 (July 1961): 117.

11. Goleman, *Emotional Intelligence*, 82–83.

12. See Daniel Akst, *Temptation: Finding Self-Control in an Age of Excess* (New York: Penguin, 2011); and Joachim de Posada and Bob Andelman, *Keep Your Eye on the Marshmallow! Gain Focus and Resilience—and Come Out Ahead in Hard Times* (New York: Penguin, 2013). A typical passage from *Keep Your Eye on the Marshmallow!* is the following: "The global economy is coming out of its worst malaise since the Great Depression of the 1930s," and a new generation of individuals are facing the need to follow the "marshmallow principles"—because "every choice [a man] makes, every

marshmallow he savors or gobbles[,] affects his wife, their two children, and even the family dog" (2). This title by de Posada and Andelman is a sequel to Joachim de Posada and Ellen Singer, *Don't Gobble the Marshmallow . . . Ever! The Secret Success in Times of Change* (New York: Penguin, 2007); and Joachim de Posada and Ellen Singer, *Don't Eat the Marshmallow . . . Yet! The Secret to Sweet Success in Work and Life* (New York: Penguin, 2005).

13. Daniel Goleman, "What Makes a Leader?," *Harvard Business Review*, November–December 1998, 93.

14. Erica Brown, *Spiritual Boredom: Rediscovering the Wonder of Judaism* (Woodstock, Vt.: Jewish Lights Publishing, 2009), 85.

15. Richard L. Peterson and Frank F. Murtha, *MarketPsych: How to Manage Fear and Build Your Investor Identity* (Hoboken, N.J.: Wiley and Sons, 2010), 132.

16. John D. Greenwood, *The Disappearance of the Social in Social Psychology* (New York: Cambridge University Press, 2004).

17. See Walter Mischel, *Personality and Assessment* (New York: John Wiley and Sons, 1968). Also see Walter Mischel, "Toward a Cognitive Social Learning Reconceptualization of Personality," *Psychological Review* 80 (July 1973): 252–83.

18. David Brooks, "Marshmallows and Public Policy," *New York Times*, May 7, 2006, A13. See also David Brooks, "The Character Factory," *New York Times*, July 31, 2014, A23.

19. Allison Davis, *Social-Class Influences upon Learning* (Cambridge, Mass.: Harvard University Press, 1948), 74.

20. Allison Davis, "The Socialization of the American Negro Child and Adolescent," *Journal of Negro Education* 8 (July 1939): 264. See also Allison Davis and Robert J. Havighurst, "Social Class and Color Differences in Child-Rearing," *American Sociological Review* 11 (December 1946): 698–710; and Michael R. Hillis, "Allison Davis and the Study of Race, Social Class, and Schooling," *Journal of Negro Education* 64 (Winter 1995): 33–41.

21. John Dollard, *Caste and Class in a Southern Town* (New Haven: Yale University Press, 1937), 425–27.

22. Martha C. Ericson, "Child-Rearing and Social Status," *American Journal of Sociology* 52 (November 1946): 192.

23. Robert J. Havighurst, "Child Development in Relation to Community Social Structure," *Child Development* 17 (March–June 1946): 87–88.

24. Arnold W. Green, "The Middle Class Male Child and Neurosis," *American Sociological Review* 11 (February 1946): 35, 37–38.

25. See Betty Friedan, *The Feminine Mystique* (1963; repr., New York: W. W. Norton, 2001), 289–93.

26. Albert K. Cohen, *Delinquent Boys: The Culture of the Gang* (New York: Free Press, 1955), 89–90, 97–99.

27. David Matza and Gresham M. Sykes, "Juvenile Delinquency and Subterranean Values," *American Sociological Review* 26 (October 1961): 716. Also see Gresham M. Sykes and David Matza, "Techniques of Neutralization," *American Sociological Review* 22 (December 1957): 664–70.

28. Eleanor E. Maccoby, Patricia K. Gibbs, and the Staff of the Laboratory of Human Development, Harvard University, "Methods of Child-Rearing in Two Social Classes," in *Readings in Child Development*, ed. William E. Martin and Celia B. Stendler (New York: Harcourt, Brace, 1954), 395.

29. Urie Bronfenbrenner, "Socialization and Social Class through Time and Space," in *Readings in Social Psychology*, ed. Eleanor E. Maccoby et al. (New York: Holt, Rinehart and Winston, 1958), 423. See Benjamin Spock, *Baby and Child Care* (New York: Pocket Books, 1957); and Robert E. Sears et al., *Patterns of Child Rearing* (Evanston: Row, Peterson, 1957), esp. 442–47.

30. William H. Sewell, "Social Class and Childhood Personality," *Sociometry* 24 (December 1961): 350.

31. Sheldon Glueck and Eleanor Glueck, "Predictive Techniques in the Prevention and Treatment

of Delinquency," in *Ventures in Criminology: Selected Recent Papers* (London: Tavistock, 1964), 92–93.

32. Sheldon Glueck and Eleanor Glueck, *Unraveling Juvenile Delinquency* (New York: Commonwealth Fund, 1950), 282.

33. Or so the Gluecks themselves announced in 1959. See their summary in Sheldon Glueck and Eleanor Glueck, *Predicting Delinquency and Crime* (Cambridge, Mass.: Harvard University Press, 1959), 127–36.

34. Charles Bennett, "Bronx Study Spots Likely Delinquents: City Study Spots Likely Bad Boys," *New York Times*, January 22, 1960, 1.

35. Lee N. Robins, "Sturdy Predictors of Adult Antisocial Behavior, Replications from Longitudinal Studies," *Psychological Medicine* 8 (1978): 611. Also see Lee N. Robins, *Deviant Children Grown Up: A Sociological and Psychiatric Study of Sociopathic Personality* (Baltimore: Williams and Wilkins, 1966).

36. Glueck and Glueck, *Predicting Delinquency and Crime*, 59.

37. See Edward L. Thorndike, *Educational Psychology, Vol. 1: The Original Nature of Man* (New York: Teachers College, Columbia University, 1913).

38. On rats as trial subjects, see Frank A. Logan, "The Role of Delay of Reinforcement in Determining Reaction Potential," *Journal of Experimental Psychology* 43 (June 1952): 393–99. On pigeons as trial subjects, see Charles B. Ferster, "Sustained Behavior under Delayed Reinforcement," *Journal of Experimental Psychology* 45 (April 1953): 218–24. A general summary of animal research on reinforcement delay can be found in Glenn Terrell, "Delayed Reinforcement Effects," in *Advances in Child Development and Behavior*, vol. 2, ed. Lewis P. Lipsitt and Charles C. Spiker (New York: Academic Press, 1965), 129–32.

39. A general summary of early research into reinforcement delay and human subjects can be found in K. Edward Renner, "Delay of Reinforcement: A Historical Review," *Psychological Bulletin* 61 (May 1964): 355–57.

40. Mischel, "Preference for Delayed Reinforcement," 57, 60.

41. Julian B. Rotter, *Social Learning and Clinical Psychology* (New York: Prentice-Hall, 1954), 434.

42. Hugh Hartshorne and Mark May, *Studies in the Nature of Character, Vol. 1: Studies in Deceit* (New York: Macmillan, 1928), 412.

43. Frank Astor, "Studies in Deceit and Self-Control," *Journal of Education* 112 (1931): 730.

44. Mischel, *Personality and Assessment*, 23.

45. Mischel and Gilligan, "Delay of Gratification," 412.

46. Walter Mischel, *The Marshmallow Test: Mastering Self-Control* (New York: Little, Brown, 2014), 74.

47. Mischel and Gilligan, "Delay of Gratification," 412.

48. Daniel Patrick Moynihan, *The Negro Family: The Case for National Action* (Washington, D.C.: Office of Policy Planning and Research, U.S. Department of Labor, March 1965), 29.

49. Daryl Michael Scott, *Contempt and Pity: Social Policy and the Image of the Damaged Black Psyche, 1880–1996* (Chapel Hill: University of North Carolina Press, 1997), 151–52.

50. See, for instance, the contributions to Lee Rainwater and William L. Yancey, eds., *The Moynihan Report and the Politics of Controversy* (Cambridge, Mass.: MIT Press, 1967).

51. Ellen Herman, *The Romance of American Psychology: Political Culture in the Age of Experts* (Berkeley: University of California Press, 1995), 205.

52. As Ryan wrote, "Moynihan was able to take a subject that had previously been confined to the Sociology Department seminar room, filled with aromatic smoke from judiciously puffed pipes, and bring it into a central position in popular American thought, creating a whole new set of group stereotypes which support the notion that Negro culture produces a weak and disorganized form of family life, which in turn is a major factor in maintaining Negro inequality." See William Ryan, *Blaming the Victim* (New York: Pantheon Books, 1971), 62.

53. In more recent years, there has been an argument that these criticisms caricatured Moynihan's actual viewpoints. See, for instance, James T. Patterson, *Freedom Is Not Enough: The Moynihan Report and America's Struggle over Black Family Life—from LBJ to Obama* (New York: Basic Books, 2010). For continued critical assessments of the Moynihan Report, see Leigh Mullings, *On Our Own Terms: Race, Class, and Gender in the Lives of African American Women* (New York: Routledge, 1997), 116–18, 161–62; and Melissa V. Harris-Perry, *Sister Citizen: Shame, Stereotypes, and Black Women in America* (New Haven: Yale University Press, 2011), 93–94.

54. Thomas F. Pettigrew, *A Profile of the Negro American* (Princeton: Van Nostrand, 1964).

55. Moynihan, *The Negro Family*, 39.

56. The only study I have found that mentions this aspect of the Moynihan Report is Alice O'Connor, *Poverty Knowledge: Social Science, Social Policy, and the Poor in Twentieth-Century U.S. History* (Princeton: Princeton University Press, 2001), 205.

57. *Manpower Report of the President* (Washington, D.C.: U.S. Government Printing Office, April 1968), 86.

58. Edward C. Banfield, *The Unheavenly City: The Nature and Future of Our Urban Crisis* (Boston: Little, Brown, 1968), 53.

59. Amitai Etzioni, "Human Beings Are Not Very Easy to Change After All," *Saturday Review*, June 3, 1972, 47.

60. O'Connor, *Poverty Knowledge*, 110–11.

61. S. M. Miller, Frank Riessman, and Arthur A. Seagull, "Poverty and Self-Indulgence: A Critique of the Non-deferred Gratification Pattern," in *Poverty in America: A Book of Readings*, ed. Louis Ferman et al. (Ann Arbor: University of Michigan Press, 1965), 286, 301, 300. The experiment cited had been conducted by psychologist Arthur A. Seagull. Also see S. M. Miller and Frank Riessman, *Social Class and Social Policy* (New York: Basic Books, 1968).

62. Walter Mischel, *Introduction to Personality* (New York: Holt, Rinehart and Winston, 1971), 155.

63. Walter Mischel, "Processes in Delay of Gratification," in *Advances in Experimental Social Psychology*, vol. 7, ed. Leonard Berkowitz (New York: Academic Press, 1974), 253–54, 250.

64. See, for instance, R. J. Herrnstein, "Some Factors Influencing Behavior in a Two-Response Situation," *Transactions of the New York Academy of Sciences* 21 (November 1958): 35–45; and R. J. Herrnstein, "Relative and Absolute Strength of Response as a Function of Frequency of Reinforcement," *Journal of the Experimental Analysis of Behavior* 4 (July 1961): 267–72.

65. James Q. Wilson and Richard J. Herrnstein, *Crime and Human Nature* (New York: Simon and Schuster, 1985), 243.

66. Walter Mischel and Ralph Metzner, "Preference for Delayed Reward as a Function of Age, Intelligence, and Length of Delay Interval," *Journal of Abnormal and Social Psychology* 64 (June 1962): 425–31.

67. See Richard Herrnstein, "Some Criminogenic Traits of Offenders," in *Crime and Public Policy*, ed. James Q. Wilson (San Francisco: Institute for Contemporary Studies, 1983), 31–52.

68. Michael R. Gottfredson and Travis Hirschi, *A General Theory of Crime* (Stanford: Stanford University Press, 1990), 89, 117, 232. Also see Travis Hirschi, *Causes of Delinquency* (Berkeley: University of California Press, 1969).

69. See the excellent collection of essays in Erich Goode, ed., *Out of Control: Assessing the General Theory of Crime* (Stanford: Stanford University Press, 2008).

70. Richard J. Herrnstein, *I.Q. in the Meritocracy* (Boston: Little, Brown, 1973). Also see Richard J. Herrnstein, "I.Q.," *Atlantic*, September 1971, 43–64.

71. Richard J. Herrnstein and Charles Murray, *The Bell Curve: Intelligence and Class Structure in American Life* (New York: Free Press, 1994), 25, 113.

72. "The children of the new upper class are the object of intense planning from the moment the woman learns she is pregnant," Murray noted. "She makes sure her nutritional intake exactly mirrors the optimal diet and takes classes (along with her husband) to prepare for a natural

childbirth — a C-section is a last resort." Also after the birth, an upper-class mother behaves perfectly: "She breast-feeds her new-born, usually to the complete exclusion of formula, and tracks the infant's growth with the appropriate length and weight charts continually. The infant is bombarded with intellectual stimulation from the moment of birth. . . . The mobile over the infant's crib and the toys with which he is provided are designed to induce every possible bit of neural growth within the child's cerebral cortex." Charles Murray, *Coming Apart: The State of White America, 1960–2010* (New York: Crown Forum, 2012), 53, 39.

73. Michael Bérubé, *Life as We Know It: A Father, a Family, and an Exceptional Child* (New York: Pantheon Books, 1996), 246.

74. Writing in 2011, David Brooks observed about Murray's contributions that they demonstrated how "a hereditary meritocratic class" successfully "passed down habits, knowledge, and cognitive traits" and thus successfully "reinforces itself through genes and strenuous cultivation generation after generation." See David Brooks, *The Social Animal: The Hidden Sources of Love, Character, and Achievement* (New York: Random House, 2011), 107.

75. Angela L. Duckworth, "The Significance of Self-Control," *Proceedings of the National Academy of Sciences of the United States of America* 108 (February 15, 2011): 2639. Also see this rhetorical query from positive psychologists Christopher Peterson and Martin E. P. Seligman: "Is there a downside to self-regulation? Thus far, we are unaware of any undesirable consequences or correlates of high self-control." There was really no such thing as "excessive self-control." Christopher Peterson and Martin E. P. Seligman, *Character Strengths and Virtues: A Handbook and Classification* (New York: Oxford University Press, 2004), 508.

76. Paul Tough, *How Children Succeed: Grit, Curiosity, and the Hidden Power of Character* (Boston: Houghton Mifflin Harcourt, 2012), xv.

77. As discussed in chapter 1, Seligman's early research involved laboratory experiments with rats and dogs.

78. Martin E. P. Seligman, *Learned Optimism* (New York: Knopf, 1991), 40–41.

79. Martin E. P. Seligman, *Authentic Happiness: Using the New Positive Psychology to Realize Your Potential for Lasting Fulfillment* (New York: Free Press, 2002), 13. For a deeply researched history of "the business of happiness," see Daniel Horowitz, *Happier? The History of a Cultural Movement That Aspired to Transform America* (New York: Oxford University Press, 2018).

80. Angela L. Duckworth and Martin E. P. Seligman, "Self-Discipline Outdoes IQ in Predicting Academic Performance of Adolescents," *Psychological Sciences* 16 (December 2005): 942.

81. Terrie E. Moffitt et al., "A Gradient of Childhood Self-Control Predicts Health, Wealth, and Public Safety," *Proceedings of the National Academy of Sciences of the United States of America* 108 (February 15, 2011): 2693.

82. B. J. Casey et al., "Behavioral and Neural Correlates of Delay of Gratification 40 Years Later," *Proceedings of the National Academy of Sciences of the United States of America* 108 (September 6, 2011): 14998–15003.

83. Marc G. Berman et al., "Dimensionality of Brain Networks Linked to Life-Long Individual Differences in Self-Control," *Nature Communications* 4 (January 22, 2013): 6.

84. There was no mention of Mischel's gratification delay experiments for instance in either Jonathan Freedman et al., *Social Psychology*, 4th ed. (Englewood Cliffs, N.J.: Prentice-Hall, 1981); or Kenneth J. Gergen and Mary M. Gergen, *Social Psychology*, 2nd ed. (New York: Springer-Verlag, 1986). Even when notice was taken, psychology textbooks interpreted these experiments "within the framework of the more complex conception of identification and modeling behavior" or in terms of the "selectivity of information processing" (that is, the ability of some children to distract themselves). See Paul F. Secord and Carl W. Backman, *Social Psychology*, 2nd ed. (New York: McGraw-Hill, 1974), 496; and Edward E. Sampson, *Social Psychology and Contemporary Society* (New York: John Wiley and Sons, 1971), 98. That these experiments might have anything significant to say about the likelihood of future life success was not immediately apparent.

85. Susan T. Fiske and Shelley E. Taylor, *Social Cognition* (Reading, Mass.: Addison-Wesley, 1984).

86. Sharon Begley with Jean Chatzky, "Moneybrain: The New Science behind Your Spending Addiction," *Newsweek*, November 7, 2011, 52–54.

87. Jeffrey Kluger, "Getting to No: The Science of Building Willpower," *Time*, March 5, 2012.

88. See David C. Funder and Jack Block, "The Role of Ego-Control, Ego-Resilience, and IQ in Delay of Gratification in Adolescence," *Journal of Personality and Social Psychology* 57 (December 1989): 1049.

89. Jack Block and Adam M. Kremen, "IQ and Ego-Resiliency: Conceptual and Empirical Connections and Separateness," *Journal of Personality and Social Psychology* 70 (February 1996): 351. Also see Jack Block, *Personality as an Affect-Processing System: Toward an Integrative Theory* (Mahwah, N.J.: Lawrence Erlbaum, 2002).

90. See, for instance, Jack H. Block et al., "Personality Antecedents of Depressive Tendencies in 18-Year-Olds: A Prospective Study," *Journal of Personality and Social Psychology* 60 (May 1991): 726–38; Darya L. Zabelina et al., "The Psychological Tradeoffs of Self-Control," *Personality and Individual Differences* 43 (August 2007): 463–73; and Janet Polivy, "The Effects of Behavioral Inhibition: Integrating Internal Cues, Cognition, Behavior, and Affect," *Psychological Inquiry* 9 (1998): 181–204.

91. Gregory Miller quoted in Alina Tugend, "Winners Never Quit? Well, Yes, They Do," *New York Times*, August 15, 2008, C5.

92. Alfie Kohn, "Why Self-Discipline Is Overrated: The (Troubling) Theory and Practice of Control from Within," *Phi Delta Kappan* 90 (November 2008): 171, 174.

93. Angela Lee Duckworth, "Self-Discipline Is Empowering," *Phi Delta Kappan* 90 (March 2009): 536.

94. Quoted in Simon Makin, "A Marshmallow in the Hand," *Scientific American Mind*, March 2013, 8.

95. Celeste Kidd et al., "Rational Snacking: Young Children's Decision-Making on the Marshmallow Task Is Moderated by Beliefs about Environmental Reliability," *Cognition* 126 (January 2013): 109–10.

96. Bruce Bower, "Delaying Gratification Is about Worldview as Much as Willpower," *Science News*, November 17, 2012, 10.

97. Paula S. Fass, "The IQ: A Cultural and Historical Framework," *American Journal of Education* 88 (August 1980): 436. For another study from this moment that analyzed the historical uses (and misuses) of IQ, see Stephen Jay Gould, *The Mismeasure of Man* (New York: W. W. Norton, 1981).

98. Goleman, *Emotional Intelligence*, 78.

99. Nancy Gibbs and Sharon E. Epperson, "The EQ Factor," *Time*, October 2, 1995.

100. The decline of the white middle class into ever-greater insecurity is acknowledged across the political spectrum — not least by *The Bell Curve*'s coauthor Charles Murray. See his *Coming Apart*.

Chapter 5

1. Hillary Rodham Clinton, *It Takes a Village* (New York: Simon and Schuster, 1996), 43, 53, 47, 49.

2. Clinton, 48–51.

3. Carla Shatz and Deborah A. Phillips quoted in "The White House Conference on Early Childhood Development and Learning: What New Brain Research Tells Us about Young Children" April 17, 1997, available at http://listserv.ed.gov/archives/edinfo/archived/msg00272.html (accessed March 13, 2018).

4. Hillary Clinton's remarks on April 17, 1997, available at https://clinton3.nara.gov/WH/New/ECDC/Remarks.html (accessed March 13, 2018).

5. On the Clinton White House's commitment to use findings from brain science to push for policies to support early childhood development, see Joe Klein, "Clintons on the Brain," *New Yorker*, March 17, 1997, 59–63.

6. Jerome Kagan, *Three Seductive Ideas* (Cambridge, Mass.: Harvard University Press, 1998), 87.

7. John T. Bruer, "Education and the Brain: A Bridge Too Far," *Educational Researcher* 26 (November 1997): 7, 15.

8. John T. Bruer, "The Brain and Child Development: Time for Some Critical Thinking," *Public Health Report* 113 (September–October 1998): 389.

9. John T. Bruer, "Brain Science, Brain Fiction," *Educational Leadership* 56 (November 1998): 16, 18. Also see John T. Bruer, *The Myth of the First Three Years: A New Understanding of Early Brain Development and Lifelong Learning* (New York: Free Press, 1999). In a posting on its website, the child advocacy organization Zero to Three acknowledged that "the neuroscience of early childhood is, in a sense, in its own infancy" and that "a child's brain is not even close to being completely wired when the third candle on the birthday cake has been blown out." But the organization also angrily struck back at Bruer, accusing him of spreading his own "misconceptions that efforts to help young children are a waste of money and time." Zero to Three speculated that it "may be Bruer's intent" to undermine "the very modest funding" that early intervention programs currently received; the group observed that Bruer "conveniently omits from his book" the "ample evidence for the value of early intervention." Zero to Three went on to caution that there were genuine "dangers" in Bruer's attacks on early childhood research, not least of which was that these attacks could help to spread the opinion that "intervening in the lives of very young children at risk for poor outcomes in school and adulthood will have little or no effect." "Zero to Three Response to *The Myth of the First Three Years*," available at http://people.ucalgary.ca/~mueller/EDPS635/ZeroThreeMyth.html (accessed March, 13, 2018).

10. Steven Pinker, *The Blank Slate: The Modern Denial of Human Nature* (New York: Penguin Books, 2002), 386–87.

11. Richard J. Herrnstein and Charles Murray, *The Bell Curve: Intelligence and Class Structure in American Life* (New York: Free Press, 1994), 298, 519.

12. Robert H. Bradley and Robert F. Corwyn, "Socioeconomic Status and Child Development," *Annual Review of Psychology* 53 (February 2002): 371.

13. Martha J. Farah et al., "Childhood Poverty: Specific Associations with Neurocognitive Development," *Brain Research* 1110 (September 19, 2006): 166, 170. Also see Kimberly G. Noble, M. Frank Norman, and Martha J. Farah, "Neurocognitive Correlates of Socioeconomic Status in Kindergarten Children," *Developmental Science* 8 (January 2005): 74–87.

14. The term "racial differences in intelligence" (along with "racial differences in IQ" and other similar phrasings) appeared repeatedly in *The Bell Curve*. Herrnstein and Murray relied on the article "Racial Differences in Intelligence" for their own analysis; they characterized its author, psychologist Richard Lynn, as "a leading scholar of racial and ethnic differences." Herrnstein and Murray, *The Bell Curve*, 272. See Richard Lynn, "Racial Differences in Intelligence: A Global Perspective," *Mankind Quarterly* 31 (Spring 1991): 255–96. Here Lynn asserted (among other things) that "the mean IQs of Negroids have invariably been found to be substantially lower than those of Caucasoids" (270). As for the journal *Mankind Quarterly*, sociologist Stephen Steinberg has labeled it "a neo-fascist journal that espouses the genetic superiority of the white race." See Stephen Steinberg, "America Again at the Crossroads," in *Theories of Race and Racism: A Reader*, ed. Les Back and John Solomos (New York: Routledge, 2000), 571. For more on Lynn, also see Charles Lane, "The Tainted Sources of 'The Bell Curve,'" *New York Review of Books*, December 1, 1994, 14–19.

15. Bob Herbert, "Throwing a Curve," *New York Times*, October 26, 1994, A27.

16. See, for instance, Adolph Reed Jr., "Looking Backward," *Nation*, November 28, 1994, 654–62; Stephen Jay Gould, "Curveball," *New Yorker*, November 28, 1994, 139, 147–48; Leon Kamin, "Behind the Curve," *Scientific American*, February 1995, 99–102; and Jacqueline Jones, "Back to the Future with *The Bell Curve*: Jim Crow, Slavery, and G," in *The Bell Curve Wars: Race, Intelligence, and the Future of America*, ed. Steven Fraser (New York: Basic Books, 1995), 80–93. Also see Claude S. Fischer et al., *Inequality by Design: Cracking the Bell Curve Myth* (Princeton: Princeton University Press, 1996). And historian Jacqueline Jones reflected that *The Bell Curve* offered a romanticized "mythology

of the American past" when "different groups literally 'knew their place' within the structure of American society." The book, she added, thus benefited by being published in a time of deep economic uncertainty: "Indeed, *The Bell Curve* is at its core a polemic about the structure of the American labor force; by ranking in one long queue all potential workers according to their 'IQ,' the authors seek to reserve the best jobs for people who score well on standardized intelligence tests." See Jacqueline Jones, *American Work: Four Centuries of Black and White Labor* (New York: Norton, 1998), 387–88.

17. Christopher Winship, "Lessons Beyond 'The Bell Curve,'" *New York Times*, November 15, 1994, A29.

18. James J. Heckman, "Lessons from the Bell Curve," *Journal of Political Economy* 103 (October 1995): 1102. In the twenty-first century, Heckman began to advocate that intervention programs for very young disadvantaged children saw optimal rates of return and were therefore excellent capital investments. See James J. Heckman, "Skill Formation and the Economics of Investing in Disadvantaged Children," *Science* 312 (June 30, 2016): 1900–1902.

19. Ulric Neisser et al., "Intelligence: Knowns and Unknowns," *American Psychologist* 51 (February 1996): 84, 86–87, 96–97.

20. Linda S. Gottfredson, "Mainstream Science on Intelligence," *Wall Street Journal*, December 13, 1994, A18.

21. Herrnstein and Murray, *The Bell Curve*, 522–23, 541.

22. Linda S. Gottfredson, "Why *g* Matters: The Complexity of Everyday Life," *Intelligence* 24 (January/February 1997): 96, 117, 123–25.

23. Gottfredson, 83, 87, 120. Also see David C. Rowe, "A Place at the Policy Table? Behavior Genetics and Estimates of Family Environmental Effects on IQ," *Intelligence* 24 (January/February 1997): 133–58.

24. Daryl Michael Scott, *Contempt and Pity: Social Policy and the Image of the Damaged Black Psyche, 1880–1996* (Chapel Hill: University of North Carolina Press, 1997), 1.

25. Linda S. Gottfredson, "Egalitarian Fiction and Collective Fraud," *Society* 31 (March/April 1994): 53, 55–56. Or as Gottfredson also added, "All that is required is for scientists to act like scientists — to demand, clearly and consistently, respect for truth and for free inquiry in their own settings, and to resist the temptation to win easy approval by endorsing a comfortable lie" (59).

26. Quoted in Dinesh D'Souza, *The End of Racism: Principles for a Multicultural Society* (New York: Free Press, 1995), 431.

27. Sharon Begley, "How to Build a Baby's Brain," *Newsweek*, Spring/Summer 1997, special ed., 28.

28. Rima Shore, *Rethinking the Brain: New Insights into Early Development* (New York: Families and Work Institute, 1997), 36.

29. J. Madeleine Nash, "Fertile Minds," *Time*, February 3, 1997.

30. Sharon Begley, "Your Child's Brain," *Newsweek*, February 19, 1996, 55.

31. Debra Viadero, "Brain Trust," *Education Week*, September 18, 1996.

32. Jennifer Lach, "Cultivating the Mind," *Newsweek*, Spring/Summer 1997, special ed., 38.

33. Ronald Kotulak, *Inside the Brain: Revolutionary Discoveries of How the Mind Works* (Kansas City, Mo.: Andrews McMeel, 1996), 7. Hillary Clinton cited Kotulak's writing on the brain in *It Takes a Village*. See Clinton, *It Takes a Village*, 47–48. And certainly Kotulak did not interpret these findings from brain science in a pessimistic light. On the contrary, he wrote that "the new findings are expected to have a revolutionary impact on early child-rearing practices and education, and provide further support for such intervention programs as Head Start." Kotulak, *Inside the Brain*, 52.

34. Julee J. Newberger, "New Brain Development Research — a Wonderful Window of Opportunity to Build Public Support for Early Childhood Education!," *Young Children* 52 (May 1997): 4–9. Ethologist Patrick Bateson had first introduced in the 1970s a concept that there existed critical or sensitive "windows of opportunity" when life experiences might most effectively influence brain development. Bateson had likened brain development to a moving trolley with "all its windows closed for the first part of the journey" — only then "at a particular moment all the windows are

thrown open and the passengers are exposed to the outside world. A little later all the windows are shut again." Although Bateson then added, "A more refined thought is that the train is divided up into compartments and the different windows open and shut at different stages during the journey." Patrick Bateson, "How Do Sensitive Periods Arise and What Are They For?," *Animal Behavior* 27 (May 1979): 471.

35. Kotulak, *Inside the Brain*, 7.

36. Donald O. Hebb, *The Organization of Behavior: A Neuropsychological Theory* (New York: Wiley, 1949), 12.

37. Marian C. Diamond, David Krech, and Mark R. Rosenzweig, "The Effects of an Enriched Environment on the Histology of the Rat Cerebral Cortex," *Journal of Comparative Neurology* 123 (August 1964): 111. Also see Mark R. Rosenzweig, David Krech, Edward L. Bennett, and Marian C. Diamond, "Effects of Environmental Complexity and Training on Brain Chemistry and Anatomy: A Replication and Extension," *Journal of Comparative and Physiological Psychology* 55 (August 1962): 429–37.

38. Edward L. Bennett, Marian C. Diamond, David Krech, and Mark R. Rosenzweig, "Chemical and Anatomical Plasticity of Brain," *Science* 146 (October 30, 1964): 610.

39. Mark R. Rosenzweig, "Environmental Complexity, Cerebral Change, and Behavior," *American Psychologist* 21 (April 1966): 321–32.

40. William T. Greenough, Fred R. Volkmar, and Janice M. Juraska, "Effects of Rearing Complexity on Dendritic Branching in Frontolateral and Temporal Cortex of the Rat," *Experimental Neurology* 41 (November 1973): 371.

41. William T. Greenough, James E. Black, and Christopher S. Wallace, "Experience and Brain Development," *Child Development* 58 (June 1987): 540. Also see William T. Greenough and James E. Black, "Induction of Brain Structure by Experience," in *Developmental Behavior Neuroscience: The Minnesota Symposium on Child Psychology*, vol. 24, ed. Megan R. Gunnar and Charles A. Nelson (Hillsdale, N.J.: Erlbaum, 1992), 155–200.

42. See, for instance, Charles A. Nelson, "The Neurobiological Bases of Early Intervention," in *Handbook of Early Childhood Intervention*, ed. Jack P. Shonkoff and Samuel J. Meisels, 2nd ed. (New York: Cambridge University Press, 2000), 204–27.

43. Ramey ominously added, "We are creating a backlog of children who have a great criminal potential." Ramey quoted in Kotulak, *Inside the Brain*, 52. Also see Frances A. Campbell and Craig T. Ramey, "Effects of Early Intervention on Intellectual and Academic Achievement: A Follow-Up Study of Children from Low-Income Families," *Child Development* 65 (April 1994): 684–98.

44. Craig T. Ramey and Sharon Landesman Ramey, "Early Intervention and Early Experience," *American Psychologist* 53 (February 1998): 109.

45. "Nurturing Development of the Brain," *New York Times*, editorial, April 28, 1997, A14.

46. Cited in Bruer, *Myth of the First Three Years*, 62.

47. Early Childhood Development Act of 1997, available at https://www.govtrack.us/congress /bills/105/s756/text/is (accessed March 13, 2018).

48. Sally S. Cohen, *Championing Child Care* (New York: Columbia University Press, 2001), 211.

49. Jack P. Shonkoff and Deborah A. Phillips, "From Neurons to Neighborhoods: The Science of Early Childhood Development: An Introduction," *Zero to Three* 21 (April/May 2001): 4.

50. Jack P. Shonkoff and Deborah A. Phillips, eds., *From Neurons to Neighborhoods: The Science of Early Childhood Development* (Washington, D.C.: National Academy Press, 2000), 7, 23–24, 41.

51. Shonkoff and Phillips, 21, 56, 194, 198, 275, 385, 391, 414–15.

52. In 2014 in a history of how *From Neurons to Neighborhoods* came into being, Shonkoff and his associates spoke of their "years-long effort to construct what came to be referred to as the 'core story of child development.'" That "early experiences in life build 'brain architecture'" and that "genes provide the basic instructions, but experiences leave a chemical 'signature' authorizing how and even whether the instructions are carried out," were two key concepts in this "core story." See

National Scientific Council on the Developing Child, *A Decade of Science Informing Policy: The Story of the National Scientific Council on the Developing Child* (Cambridge, Mass.: Center on the Developing Child, Harvard University, 2014), 6–7.

53. Shonkoff and Phillips, *From Neurons to Neighborhoods*, 54, 391. That a developing brain was an architectural work in progress was a metaphor that long predated *From Neurons to Neighborhoods*. However, these earlier examples chiefly used architectural language to discuss brain development in *animals*. See the landmark research of Nobel Prize winners David H. Hubel and Torsten N. Wiesel on the developing visual cortex of cats. David H. Hubel and Torsten N. Wiesel, "Receptive Fields, Binocular Interaction and Functional Architecture in the Cat's Visual Cortex," *Journal of Physiology* 160 (January 1962): 106–54. In their chapter "The Developing Brain," Shonkoff and Phillips cited Hubel and Wiesel's research with kittens, stating that "in both humans and animals, the effects of experience on [neural] systems — normal or abnormal — become increasingly irreversible over time." Shonkoff and Phillips, *From Neurons to Neighborhoods*, 189.

54. *The Science of Early Childhood Development: Closing the Gap between What We Know and What We Do* (Cambridge, Mass.: National Scientific Council on the Developing Child, 2007), 1.

55. Quoted in David Bornstein, "Protecting Children from Toxic Stress," *New York Times*, October 30, 2013, SR4.

56. Madeline Ostrander, "What Poverty Does to the Young Brain," *New Yorker*, June 4, 2015.

57. In fact, these arguments also remain remarkably tenuous. "Not only is SES a broad and poorly defined variable," observed the journal *Developmental Cognitive Neuroscience* in 2016, "but it is linked to many other factors that may confound the interpretability of potential results." The journal added, "Of course, the scientific evidence to date is far from clear." See Monica E. Ellwood-Lowe, Matthew D. Sacchet, and Ian H. Gotlib, "The Application of Neuroimaging to Social Inequity and Language Disparity: A Cautionary Examination," *Developmental Cognitive Neuroscience* 22 (December 2016): 1–8.

58. Susan B. Neuman, "Changing the Odds," *Educational Leadership* 65 (October 2007): 17, 21.

59. Herrnstein and Murray, *The Bell Curve*, 508, 519.

60. Herrnstein and Murray wrote that "there is no reason to think that any realistically improved version of Head Start, with its thousands of centers and millions of participants, can add much to cognitive functioning." They underscored that the cognitive benefits accrued to poor children in even the best preschool programs "fall short of statistical significance," bluntly adding that these "small" effects can "hardly justify investing billions of dollars in run-of-the-mill Head Start programs." Herrnstein and Murray, 405, 414, 435.

61. Herrnstein and Murray, 298, 311.

62. Indeed, it is precisely the existence of these "agnostic" and environment-attuned remarks that have served very well for defenders of Murray in the recent round of dispute over his theories. For example, see George F. Will, "The Liberals Who Loved Eugenics," *Washington Post*, March 8, 2017. Also see Katharine Q. Seelye, "Protesters Disrupt Speech by 'Bell Curve' Author at Vermont College," *New York Times*, March 3, 2017.

63. Herrnstein and Murray, *The Bell Curve*, 519.

64. Race went unmentioned in this discussion of the "intellectually gifted," but it did not go unnoticed. Psychometrician Robert J. Sternberg wrote that the model Herrnstein and Murray proffered made sense only within a remarkably "narrow conception of intelligence and giftedness" that "excludes members of many ethnic groups and favors especially those who have traditionally done well on conventional tests of intelligence." See Robert J. Sternberg et al., "Return Gift to Sender: A Review of *The Bell Curve* by Richard Herrnstein and Charles Murray," *Gifted Child Quarterly* 39 (Summer 1995): 177. Educational psychologist Donna Y. Ford was just as distressed by a proposal that seemed race-neutral but in reality merely obscured the long-standing minority underrepresentation in gifted education programs. See Donna Y. Ford, "The Underrepresentation of Minority Students in Gifted Education: Problems and Promises in Recruitment and Retention," *Journal of Special Education* 32 (Spring 1998): 4–14.

65. Herrnstein and Murray, *The Bell Curve*, 434, 442.

66. Nicholas Colangelo, Susan G. Assouline, and Miraca U. M. Gross, *A Nation Deceived: How Schools Hold Back America's Brightest Students*, vol. 1 (Iowa City: Connie Belin and Jacqueline N. Blank International Center for Gifted Education and Talent Development, University of Iowa, 2004), 1, 40.

67. Dona J. Matthews and Joanne F. Foster, "Mystery to Mastery: Shifting Paradigms in Gifted Education," *Roeper Review* 28 (Winter 2006): 64, 66, 68.

68. Sharlene D. Newman, "Neural Bases of Giftedness," in *Critical Issues and Practices in Gifted Education*, ed. Jonathan Plucker and Carolyn Callahan (Waco, Tex.: Prufrock Press, 2008), 458.

69. Charles Murray, *Real Education: Four Simple Truths for Bringing America's Schools Back to Reality* (New York: Crown Forum, 2008), 12–13, 50.

70. Rena F. Subotnik, Paula Olszewski-Kubilius, and Frank C. Worrell, "Rethinking Giftedness and Gifted Education: A Proposed Direction Forward Based on Psychological Science," *Psychological Science in the Public Interest* 12 (January 2011): 7–8.

71. Helen Neville, Courtney Stevens, Eric Pakulak, and Theodore A. Bell, "Commentary: Neurocognitive Consequences of Socioeconomic Disparities," *Developmental Science* 16 (September 2013): 709–10.

72. Kimberly G. Noble et al., "Socioeconomic Disparities in Neurocognitive Development in the First Two Years of Life," *Developmental Psychobiology* 57 (July 2015): 548.

73. Nicole L. Hair et al., "Association of Child Poverty, Brain Development, and Academic Achievement," *JAMA Pediatrics* 169 (September 2015): 828. Emphasis added.

74. Kimberley G. Noble, "How Poverty Affects Children's Brains," *Washington Post*, October 2, 2015.

75. Sara B. Johnson, Jenna L. Riis, and Kimberly G. Noble, "State of the Art Review: Poverty and the Developing Brain," *Pediatrics* 137 (April 2016): 11.

76. Ann Hulbert, *Raising America: Experts, Parents, and a Century of Advice about Children* (New York: Random House, 2003), 315.

77. Daniel A. Hackman and Martha J. Farah, "Socioeconomic Status and the Developing Brain," *Trends in Cognitive Sciences* 13 (February 2009): 69–71.

78. Amedeo D'Angiulli, Sebastian J. Lipina, and Alice Olesinska, "Explicit and Implicit Issues in the Developmental Cognitive Neuroscience of Social Inequality," *Frontiers in Human Neuroscience* 6 (September 2012): 11.

79. Kimberly G. Noble et al., "Family Income, Parental Education and Brain Structure in Children and Adolescents," *Nature Neuroscience* 18 (May 2015): 773, 778.

80. Johnson, Riis, and Noble, "State of the Art Review," 11.

81. "Remarks by the President in the State of the Union Address," February 12, 2013, at https://obamawhitehouse.archives.gov/the-press-office/2013/02/12/remarks-president-state-union-address (accessed March 13, 2018).

82. Dylan Matthews, "Read: Obama's Pre-K plan," *Washington Post*, February 14, 2013, at https://www.washingtonpost.com/news/wonk/wp/2013/02/14/read-obamas-pre-k-plan/?utm_term=.7463a8dbc4f3 (accessed March 13, 2018). Also see "The President's Plan for a Strong Middle Class and a Strong America," February 12, 2013, at https://obamawhitehouse.archives.gov/sites/default/files/uploads/sotu_2013_blueprint_embargo.pdf (accessed June 16, 2017).

83. "Preschool for All: Summary of Early Learning in the FY 2014 Presidential Budget Request," available at http://ffyf.org/resources/preschool-for-all-summary-of-early-learning-in-the-fy-2014-presidential-budget-request/ (accessed March 13, 2018).

84. Quoted in Des Griffin, *Education Reform: The Unwinding of Intelligence and Creativity* (New York: Springer, 2014), 93.

85. *2014 Preschool Development Grants* (Washington, D.C.: U.S. Department of Education, August 2014), 3.

86. U.S. Department of Education, "Fact Sheet: 100,000 Children from Low- and Moderate-Income

Families Could Lose Access to High-Quality Preschool under the 2016 House and Senate Spend-ing Bills," August 17, 2015, https://www.ed.gov/news/press-releases/fact-sheet-100000-children-low-and-moderate-income-families-could-lose-access-high-quality-preschool-under-2016-house-and-senate-spending-bills (accessed March 13, 2018).

87. Alyson Klein, "Obama's Complex Legacy on K-12: Bold Achievements, Fierce Blowback," *Education Week*, January 11, 2017, 16.

88. U.S. Department of Education and U.S. Department of Health and Human Services, "Col-laboration and Coordination of the Maternal, Infant, and Early Childhood Home Visiting Program and the Individuals with Disabilities Education Act Part C Programs," January 19, 2017, 1-2, avail-able at https://www2.ed.gov/about/inits/ed/earlylearning/files/ed-hhs-miechv-partc-guidance.pdf (accessed March 13, 2018).

89. Education Secretary Betsy DeVos did seek to profit from neuroscientific quackery through her involvement (and investment) in Neurocore, a company founded "to help children improve academic scores and lessen the need for medication to treat certain ailments." See Sheri Fine, Steve Eder, and Matthew Goldstein, "DeVos Invests in a Therapy under Doubt," *New York Times*, Janu-ary 31, 2017.

90. Catherine Y. Kim, Daniel J. Losen, and Damon T. Hewitt, *The School-to-Prison Pipeline: Structuring Legal Reform* (New York: New York University Press, 2010), 3, 113. In 2014 the Obama administration issued a "guidance document" to reduce discriminatory practices that resulted in significantly disproportionate numbers of minority students' suspensions and expulsions. In 2018 the Trump administration signaled its intention to undo the Obama-era guidance on school disci-pline, claiming that because suspensions were down, many public schools were less safe as a result. See Erica L. Green, "Trump Finds Unlikely Culprit in School Shootings," *New York Times*, March 14, 2018, A1, A10. Also see Max C. Eden, "On School Discipline, Fix the Problem, Not the Statistics," *National Review*, November 13, 2017, available at https://www.nationalreview.com/2017/11/school-discipline-federal-rules-not-helping/ (accessed March 15, 2018); and Susan Berry, "Education Secretary Betsy DeVos Urged to Scrap Obama-Era 'Black Lives Matter' School Discipline Policy," *Breitbart News*, January 2, 2018, available at http://www.breitbart.com/big-government/2018/01/02/education-secretary-betsy-devos-urged-scrap-obama-era-black-lives-matter-school-discipline-policy/ (accessed March 15, 2018).

91. Joseph Neff, Ann Doss Helms, and David Raynor, "Why Have Thousands of Smart, Low-Income NC Students Been Excluded from Advanced Classes?," *Raleigh News and Observer*, May 18, 2017, available at http://www.newsobserver.com/news/local/education/article149942987.html (accessed October 8, 2017). The investigative study appeared on the same date in the *Charlotte Observer*.

92. Michael D. Cook and William N. Evans, "Families or Schools? Explaining the Convergence in White and Black Academic Performance," *Journal of Labor Economics* 18 (October 2000): 729.

93. See Jonathan Kozol, *The Shame of the Nation: The Restoration of Apartheid Schooling in America* (New York: Crown, 2005). As Kozol wrote that same year in *Harper's*, "Schools that were already deeply segregated twenty-five or thirty years ago are no less segregated now, while thousands of other schools around the country that had been integrated either voluntarily or by the force of law have since been rapidly resegregating." Jonathan Kozol, "Still Separate, Still Unequal," *Harper's*, September 2005, 41.

94. Chief Justice John Roberts's opinion in *Parents Involved in Community Schools v. Seattle School District No. 1* (2007), available at https://www.law.cornell.edu/supremecourt/text/551/701 (accessed October 20, 2017).

95. Justice Stevens wrote, "There is cruel irony in The Chief Justice's reliance on our decision in *Brown v. Board of Education*." About Roberts's opinion, Stevens continued, "The first sentence in the concluding paragraph of his opinion states: 'Before *Brown*, schoolchildren were told where they could and could not go to school based on the color of their skin.' This sentence reminds me

of Anatole France's observation: '[T]he majestic equality of the la[w], forbid[s] rich and poor alike to sleep under bridges, to beg in the streets, and to steal their bread.' The Chief Justice fails to note that it was only black schoolchildren who were so ordered; indeed, the history books do not tell stories of the white children struggling to attend black schools. In this and other ways, The Chief Justice rewrites the history of one of this Court's most important decisions." *Parents Involved in Community Schools v. Seattle School District No. 1.*

96. Richard Rothstein, *The Color of Law: A Forgotten History of How Our Government Segregated America* (New York: Liveright, 2017), 215.

97. See Sean F. Reardon, Elena Tej Grewal, Demetra Kalogrides, and Erica Greenberg, "*Brown* Fades: The End of Court-Ordered School Desegregation and the Resegregation of American Public Schools," *Journal of Policy Analysis and Management* 31 (Fall 2012): 876–904.

98. Herrnstein and Murray, *The Bell Curve*, 509.

Afterword

1. Stephen Jay Gould, *The Mismeasure of Man*, rev. ed. (New York: W. W. Norton, 1996), 28, 367. For fuller discussion of the histories of intelligence testing outlined here, also see Leila Zenderland, *Measuring Minds: Henry Herbert Goddard and the Origins of American Intelligence Testing* (New York: Cambridge University Press, 1998); Thomas C. Leonard, *Illiberal Reformers: Race, Eugenics and American Economics in the Progressive Era* (Princeton: Princeton University Press, 2016); and Ken Richardson, *Genes, Brains, and Human Potential: The Science and Ideology of Intelligence* (New York: Columbia University Press, 2017).

INDEX

Abecedarian Project, 140

affirmative action, 75, 147, 157

African American children: and ADD/ADHD, 75, 77; and classroom environment, 1–3, 10, 26–28, 33–34, 37, 40, 42, 63; and doll tests, 3; and gifted education, 165–66; and home environment, 34, 36, 46, 116, 123–24; and IQ, 4–5, 12, 39, 46–47, 81, 93–94, 99–102, 107, 140, 142–48, 156; and juvenile justice system, 165, 208n90; and locus of control, 30, 35, 38; and mild mental retardation, 10, 59, 71; and minimal brain dysfunction, 62–66; and right-brain dominance, 11, 82, 91–97, 100–101, 106–8; and self-control, 114–16, 121; and stimulant medications, 11, 49–57, 69–71

Aggression (Bandura), 133

Alexander, Michelle, 14

American Association for the Advancement of Science, 96

American Journal of Psychiatry, 54, 104

American Journal of Psychology, 67

American Orthopsychiatric Association, 53

American Psychological Association, 4, 145

Archives of General Psychiatry, 89

attention deficit disorder (ADD), 73–76

attention-deficit/hyperactivity disorder (ADHD), 74–75, 77. *See also* hyperactivity; mild mental retardation; minimal brain dysfunction

Ausubel, David P., 19–20

Authentic Happiness (Seligman), 131

Baby and Child Care (Spock), 118

Baldwin, James, 31

Bandura, Albert, 26, 133

Banfield, Edward, 124

Banks, William B., 70

Banneker Project, 28–29, 42

Bateson, Patrick, 204–5n34

Battle, Esther, 30, 35

Beatty, Barbara, 43

Begley, Sharon, 149

Bell, Derrick, 14

Bell Curve, The (Herrnstein and Murray), 7, 12, 102, 109, 142–49, 156–58, 160, 167; and attention to environment, 157, 206n62; and cognitive stratification, 6, 128–29; and critiques of, 129, 139–40, 144, 151–52, 171

Bennett, Edward L., 150

Benzedrine. *See* stimulant medications

Bérubé, Michael, 129

Bettelheim, Bruno, 45, 69

Bickel, Alexander, 37

bilingualism, 91, 105

Binet, Alfred, 169–70

biofeedback, 11, 89–90, 192n31

biologism: and biological determinism, 4–8, 11–12, 27, 38, 43, 93, 115, 148–49, 170–71; and liberal discomfort with, 47, 64, 66, 161–63, 129; and liberal uses of biology, 95, 139–40, 148–56, 160–64

Block, Jack, 134

Bloom, Benjamin S., 19–20

Bogen, Joseph E., 83, 85–89, 91–95, 97, 100, 102

Boykin, A. Wade, 70

Bradley, Charles, 57

Brain Research, 142

Brain Revolution, The (Ferguson), 89

brain scans. *See* functional magnetic resonance imaging; positron-emission tomography

Brazelton, T. Berry, 69

Briggs v. Elliot, 2

Brigham, Carl, 170

Broca, Paul, 82

Bronfenbrenner, Urie, 62–63, 66, 187n42

Brooks, David, 114

Brown, Roger, 26

Brown v. Board of Education, 1–4, 6–8, 12, 27, 142, 159; and Coleman Report, 32–33, 35–36, 39; legacies of, 1, 13–15, 39, 166–67, 177n25, 177n29, 177n32, 208–9n95

Bruer, John T., 141, 203n9
Budzynski, Thomas H., 89–90

Caine, Geoffrey, 105
Caine, Renate, 105
Centers for Disease Control and Prevention, 74, 153
Central Intelligence Agency (CIA), 20
cerebral commissurotomy, 82–83
cerebral palsy, 59
Chambers, Ernest W., 50–51
Character Education Inquiry, 122. *See also* psychological experiments: honesty vs. cheating
Child Development, 91, 151
Chrisjohn, Roland, 105
Christian Science Monitor, 69, 73
Ciba (pharmaceutical manufacturer), 67–68
Civil Rights Act of 1964, 20, 32
Clark, Kenneth, 3, 27–29
Clark, Mamie, 3
class: as coded reference for race, 7, 62, 114, 124–25; and correlation with intelligence, 6, 128–29, 142–43, 145, 160, 170–71; and divisions within a racial grouping, 7, 14–15, 30, 57, 115–16, 177n32; and historic shift away from concern with economic injustice, 3, 8, 40, 115, 171, 176–77n22; illiberal emphasis on, 18, 37, 146; in intersection with race, 25, 47, 49–50, 70–71, 74–77, 99–100, 131, 134, 146, 165; and loss of middle-class security, 137; as more consequential than race, 22, 30, 37, 56, 66, 91–96, 114–16; and remediating poverty as best educational practice, 40, 115, 162–63; and self-control, 7, 11–12, 109–10, 114–19, 126–27, 130–31, 134–37. *See also* poverty: consequences for brain; poverty: consequences for character
Clements, Sam D., 58–60, 62, 75
Clinton, Bill, 140–41, 161
Clinton, Hillary, 139–41, 148, 151, 156, 161
cognitive deficit theory, 128
cognitive elite, 128–29, 142, 146, 158
cognitive hypothesis, 130
Cognitive Styles and the Social Order (TenHouten), 91–93
Cohen, Albert K., 117
Coleman, James S., 32–33, 35, 37
Coleman Report, 10, 32–40

Color of Law, The (Rothstein), 167, 176–77n22
Committee on Integrating the Science of Early Childhood Development, 153
compensatory education, 28–29, 33–38, 59–60, 62, 77; defunding of, 18, 35–36, 42, 164, 206n60; liberal opposition to, 38, 41–42; and perceived lack of effectiveness, 10, 22, 39, 65, 93; as worthy of investment, 8–9, 12, 19–20, 37–38, 139–42, 151–52. *See also* government funding for education
Conners, C. Keith, 53–57, 74
Conrad, Peter, 68
Corballis, Michael C., 103
Crime and Human Nature (Wilson and Herrnstein), 127
criminology, 110, 119–20, 126–28
Crisis in Black and White (Silberman), 26
Cronbach, Lee J., 42

Dark Ghetto (Clark), 29
Darley, John, 26
Davidson, Richard J., 192n34
Davis, Allison, 19–20, 114–15, 118–19, 135
delinquency, juvenile, 61, 64–65, 111, 117–20, 124
Delinquent Boys (Cohen), 117
Demonstration Guidance Project. *See* Higher Horizons Program for Underprivileged Children
Denhoff, Eric, 57–58, 67
Department of Education, 153, 164
Department of Health, Education, and Welfare, 58, 60
Department of Health and Human Services, 153, 164
Department of Labor, 123
desegregation of schools, 1–4, 10, 22, 33–34, 36–39, 63, 142, 166–67; and loss of African American role models, 10, 33. *See also* resegregation of schools
Developmental Psychobiology, 160–61
Developmental Science, 160
Deviant Children Grown Up (Robins), 120
Dexedrine. *See* stimulant medications
Diagnostic and Statistical Manual of Mental Disorders (DSM), 73–74
Diamond, Marian C., 150, 152
Divoky, Diane, 71–72
Dollard, John, 115–16, 135

Dragons of Eden, The (Sagan), 80–81
Drawing on the Right Side of the Brain
    (Edwards), 81, 105
Duckworth, Angela, 130–31, 135
Dweck, Carol S., 45, 184n108
dyslexia, 99

Early Childhood Development Act of 1997,
    152
Early Head Start, 163–64
early intervention, 12, 17–20, 151
Edelman, Marian Wright, 70
Educability and Group Differences (Jensen),
    66–67
Educational Leadership, 157
education reform, 7, 9, 37, 47, 82, 115; and
    neuroplasticity, 152–56; and self-discipline,
    130–31; and split-brain theory, 95–98, 105–8.
    See also Coleman Report; compensatory
    education; gifted education
Education Week, 149, 164
Edwards, Betty, 81, 105
Eisenberg, Leon, 53–57, 70
electroencephalogram (EEG), 89–91, 192n31,
    192n34
Elementary and Secondary Education Act,
    32–33, 39, 41–42, 47
Elementary School Journal, 25
Ellison, Ralph, 31
emotional intelligence. *See* psychological
    theories: emotional intelligence
Emotional Intelligence (Goleman), 12, 109–10,
    132
enrichment programming. *See* compensatory
    education; gifted education
environmentalism: and enmeshment with
    biologism, 8, 64, 66, 93–95, 100–101, 108,
    139–40, 153–56; and illiberal uses of, 17–18,
    145–46, 157–60; and liberal uses of, 2–7, 10,
    12, 19, 26–27, 93, 141–42, 150–56, 161–64
epilepsy, 59, 83
Equality of Educational Opportunity. See
    Coleman Report
Ericson, Martha C., 116, 118–19
Etzioni, Amitai, 125
evolutionary theory, 83–84, 133, 196n88

Farah, Martha J., 161–62
Fass, Paula, 136

father, role of, 7, 36, 39, 116, 146, 187n42; and
    children's impulse control, 111, 121, 124, 134
Fausto-Sterling, Anne, 105
federal funding for education. *See* government
    funding for education
Feingold, Ben F., 73
Feingold diet, 73
Feminine Mystique, The (Friedan), 117
Ferguson, Marilyn, 89
Fiske, Susan, 132
Forbes, 74
Fox, Nathan, 192n34
Friedan, Betty, 117
From Neurons to Neighborhoods (Shonkoff
    and Phillips), 12–13, 152–56, 159, 205n52,
    206n53
Frontiers in Human Neuroscience, 162
functional magnetic resonance imaging
    (fMRI), 132–33

g (general intelligence), 145, 147–48. *See also*
    intelligence quotient
Gage, Nathaniel L., 26
Galin, David, 89–90, 96–97, 100, 192n31
Galton, Francis, 170
Gantner, Robert L., 63
Gardner, Howard, 103, 109
Garrett, Henry E., 4
Gazzaniga, Michael S., 83, 88–89, 97, 103
Geer, James H., 45
General Theory of Crime, A (Gottfredson and
    Hirschi), 128
genetics, 10, 46–47, 108, 154–56, 201n74, 205–
    6n52; and Herrnstein and Murray, 6, 128–
    29, 140, 142, 144–46, 151–52, 157–58, 203n14;
    and Jensen, 43, 67, 81, 93–94, 96
Geschwind, Norman, 98
gifted education, 13, 157–60; and exclusion of
    children of color, 165–66; and purchase of
    gifted diagnosis, 166
Gilligan, Carol, 122–23
Glueck, Eleanor, 119–20, 127
Glueck, Sheldon, 119–20, 127
Goddard, Henry H., 169–70
Goleman, Daniel, 12, 103, 109–12, 132, 136, 139
Gordon, Edmund, 39
Gottfredson, Linda S., 145–48
Gottfredson, Michael R., 128
Gould, Stephen Jay, 169, 171

government funding for education, 17, 33–34, 59; criticisms of, 7, 12–13, 15, 18, 35, 41–42, 144, 171, 206n60; defenses of, 149, 151–52, 163–64; and reallocation to gifted, 13, 158–60

Grant, Gerald, 37

gratification delay. *See* psychological theories: gratification delay

Green, Arnold W., 116–18

Greenough, William T., 150–53

Grinspoon, Lester, 68, 73

grit scale, 130. *See also* psychological theories: grit

Grotberg, Edith H., 63

Guinier, Lani, 14, 177n32

Gunnar, Megan, 153

Hackman, Daniel A., 161–62

Hall, G. Stanley, 170

Harrington, Anne, 107

Harris, Maryl J., 49

Harvard Business Review, 79, 112

Harvard Educational Review, 5

Havighurst, Robert J., 116, 118–19

Head Start, 9, 12, 33, 39, 55, 59, 77, 151; and rationales for defunding, 17–18, 164, 182–83n83, 206n60

Hebb, Donald O., 150, 152, 178n5

Heckman, James J., 144–45

Helplessness (Seligman), 46–47

Herbert, Bob, 144

hereditarianism. *See* biologism

Heritage Foundation, 75

Herman, Ellen, 3, 123

Herrnstein, Richard J., 109, 131, 142–48, 152, 155, 157–58, 167; and cognitive stratification, 6, 129, 142; and criminology, 12, 127–28; and critiques of, 139, 144, 151, 171; and reliance on Jensen, 102; and research with pigeons, 127

Hess, Robert, 19–20

Hewitt, Damon T., 165

Higher Horizons Program for Underprivileged Children, 28, 39

Hirschi, Travis, 128

Hobson v. Hansen, 41

Holt, John, 26

How Children Fail (Holt), 26

How Children Succeed (Tough), 130

Hubel, David H., 206n53

Hughlings Jackson, John, 82, 86

Hulbert, Ann, 161

Human Emotions (Izard), 133

Hunt, Joseph McVicker, 19–20, 38

hyperactivity, 7, 10, 49–52, 56–58, 60–64, 67–74, 76–77, 187n40

I Am Your Child, 149

I Ching, 89

Identifying Hyperactive Children (Conrad), 68

impulse control. *See* self-control; class: and self-control

In Schools We Trust (Meier), 40

Inside the Brain (Kotulak), 150

Institute of Medicine, 152

Integrated Education, 40

intelligence quotient (IQ), 5–6, 9, 13, 24–25, 27–28, 140, 142, 155–56, 158; and challenges to value of metric, 8, 12, 100–102, 109, 129–31; and correlation with locus of control, 30; and correlation with race, 4–6, 10–11, 93–95, 99–101, 144, 147–48, 169–71, 178n5, 203n14; and correlation with socioeconomic status, 6, 81, 128–29, 142, 145; and helplessness, 46–47; history of, 136, 169–71; as largely immutable, 5, 19, 39, 42, 81, 93, 144, 156, 170–71; and racial bias, 82, 93–102, 145–48. *See also* plasticity of brain; testing, intelligence

Intelligent Testing with the WISC-R (Kaufman), 100

International Journal of Neuroscience, 91

interpersonal expectancy effect. *See* psychological theories: interpersonal expectancy effect

It Takes a Village (Clinton), 139, 148

Izard, Carroll E., 133

Jablonsky, Adelaide, 39

Jackson, Gerald Gregory, 96

Jackson, John P., Jr., 3

Jacobson, Lenore, 22–26, 28, 40–43

JAMA Pediatrics, 161

Jaynes, Julian, 80, 104

Jensen, Arthur R., 5, 21, 38–39, 43, 47, 66–67, 81, 93–96, 102

Jim Crow, 1, 14,

Johnson, Lyndon B., 9, 17, 32–34, 41

Johnson-Reed Immigration Act of 1924, 170

Journal of Comparative and Physiological Psychology, 20

Journal of Education, 122

Journal of Educational Psychology, 42

Journal of Experimental Psychology, 20

Journal of Learning Disabilities, 63–65, 71, 99
Journal of Negro Education, 39
Journal of School Psychology, 103
Jung, Carl, 85

Kagan, Jerome, 141
Kaiser, David, 107
Kaufman, Alan S., 11, 99–102
Kaufman, Nadeen, 101–2
Kaufman Assessment Battery for Children
    (K-ABC), 101–2
Kerry, John, 152
Kidd, Celeste, 135–36
Kim, Catherine Y., 165
King, Melvin H., 70
Klarman, Michael, 14
Kohn, Alfie, 135
Kotulak, Ronald, 150
Krech, David, 150, 152

Ladies' Home Journal, 69
Latané, Bibb, 26
Latino children, 11, 24–25, 28, 52, 69–71, 82,
    165
Laufer, Maurice, 57–58, 67
Lawrence, Vint, 76
learned helplessness. *See* psychological
    theories: learned helplessness
Learned Optimism (Seligman), 130
learning disability, 7, 22, 98–99, 159, 169; as
    distinct from retardation, 71, 76. *See also*
    mild mental retardation; minimal brain
    dysfunction
Learning Disability Quarterly, 71
LeDoux, Joseph E., 103
Lefcourt, Herbert, 31–32, 35, 45
left-brain thinking. *See* psychological theories:
    split-brain
Levy, Jerre, 83–84, 105
Lichtenberger, Elizabeth O., 102
Lieberman, Alicia F., 153
locus of control. *See* psychological theories:
    locus of control
Lorenz, Lee, 104
Losen, Daniel J., 165

Maier, Steven, 44–45
Making Connections (Caine and Caine), 105
Manpower Report of the President, 124
marshmallow test. *See* psychological
    experiments: marshmallow test

Marshmallow Test, The (Mischel), 113
Mayer, John D., 109–10
McCall's, 69
McCray, Patrick, 107
meditation, 11, 80, 88–89
Meier, Deborah W., 40
Metaphoric Mind, The (Samples), 96
methylphenidate. *See* Ritalin
mild mental retardation, 10, 60; development
    of concept, 59
Milgram, Stanley, 26
Miller, Zell, 106
minimal brain dysfunction (MBD), 10–11, 50,
    52, 56, 58–60, 67–68, 72–73, 75, 77; and
    African American children, 62–66, 70;
    development of concept, 58–59
Mintzberg, Henry, 79
Miró, Joan, 85
Mischel, Walter, 11–12, 110–13, 121–24, 126–
    28, 130, 132–33, 135
Mismeasure of Man, The (Gould), 171
Montessori, Maria, 19, 38
Moore, Henry, 85
mother, role of, 28, 36, 46, 106, 116–18, 142,
    157, 200–201n72; and hyperactive children,
    50, 56, 58, 61–62, 69
Moynihan, Daniel Patrick, 12, 21, 35–36, 39, 123
Moynihan Report, 12, 36, 123–24; and promo-
    tion of stereotypes, 199n52, 200n53
Mozart effect, 106
Murray, Charles A., 109, 131, 142–48, 152, 155,
    157–58, 167; and brain plasticity, 159–60;
    and cognitive stratification, 6, 129, 142;
    and critiques of, 139, 144, 151, 171; and ideal
    mothering, 200–201n72; and reliance on
    Jensen, 102
Myers, Ronald E., 82
My Fair Lady (Lerner and Loewe), 23
Myth of the Hyperactive Child, The (Schrag
    and Divoky), 71–72

National Academy of Sciences' Board on Chil-
    dren, Youth, and Families, 152
National Association for the Advancement of
    Colored People Legal Defense Fund, 3
National Education Association of the United
    States, 98
National Institute of Mental Health, 71, 73, 153
National Research Council, 152
National Scientific Council on the Developing
    Child, 156

Nation Deceived, A (Templeton Foundation), 158–59

Native American children, 11, 82, 91–93, 95, 105

Nature Neuroscience, 162

Nature of Human Consciousness, The (Ornstein), 89

nature vs. nurture, 64, 145, 152–56. *See also* biologism; environmentalism

Nazism, 13, 45, 171

Nebes, Robert D., 103

Negro Family, The (Moynihan). *See* Moynihan Report

neuroeducation, 106

neuroplasticity. *See* plasticity of brain

neuropsychology, 7, 11, 79, 89–92, 105–8

neuroscience, 12–15, 130, 132–33; and early childhood development, 139–43, 149–56, 162–65; and gifted education, 159–60; history of, 9; and shift from lateralization to localization, 106–8

New England Journal of Medicine, 62

New Jim Crow, The (Alexander), 14

New Republic, 69, 75–76

Newsweek, 50, 133, 149

New Yorker, 102, 104, 156

New York Times, 37, 40, 114, 120, 135, 144, 151, 156; and split-brain research, 88, 98; and stimulant medications, 67, 74

New York Times Magazine, 79, 88

Nixon, Richard M., 9, 17–18, 20–22, 29, 35, 37, 39, 45–46

Noble, Kimberly G., 161

Obama, Barack, 163–64

Oberst, Byron B., 49–50

O'Connor, Alice, 125

Origins of Consciousness in the Breakdown of the Bicameral Mind, The (Jaynes), 80

Ornstein, Robert, 88–90, 96–97, 192n31

Overmier, J. Bruce, 44–45

Parents Involved in Community Schools v. Seattle School District No. 1, 166–67, 208–9n95

Parents Magazine, 69

Pearl, Arthur, 27, 29

Pediatrics, 161

Peters, Michael, 105

Phillips, Deborah A., 140, 153–54

Pines, Maya, 88, 96

Pinker, Steven, 141

plasticity of brain, 46, 90–91, 108, 141, 169; as double-edged sword, 155; and illiberal uses of, 13, 35, 143, 158–60; and liberal uses of, 7–8, 12, 19–20, 29, 106, 140–43, 148–56, 164; and skepticism about, 10, 39, 43, 47, 144

plasticity of character, 134, 136–37, 145

Plessy, Homer, 1–2

Plessy v. Ferguson, 1–2, 175n2, 175n5

Policy Review, 75

positive psychology, 20, 130–31. *See also* psychological theories: learned helplessness; psychological theories: learned optimism

positron-emission tomography (PET), 149

poverty: consequences for brain, 8, 12–13, 59, 93, 141–44, 149, 151–58, 160–64; consequences for character, 45–46, 114–18, 124, 126, 134; and diagnosis of minimal brain dysfunction, 10, 62–64, 66; and learned helplessness, 18, 45–46. *See also* neuroscience: and early childhood development; psychological theories: gratification delay; psychological theories: locus of control

Power of their Ideas, The (Meier), 40

Preschool Development Grants, 163–64

President's Council on Bioethics, 74

Profile of the Negro American, A (Pettigrew), 124

psychological experiments, 6, 9; and limits of laboratory setting, 23–24, 134, 137; and use of deception, 122–23, 126

—with animals, 21, 23–24, 43–45, 120, 127, 130, 150, 179n22

—bystander apathy, 26

—dichotic listening, 92, 104–5

—doll tests, 3, 27

—experimenter bias, 23–24, 179n22

—hemispheric activation, 89–92, 97–98

—honesty vs. cheating, 122–23, 135

—marshmallow test, 11, 110–14, 127, 130–33, 135–36; history of, 110–11, 121–23, 197n6, 197–98n12

—obedience to authority, 26

—reinforcement schedule, 29, 111, 120–21, 126–27, 130, 199n38

—social modeling, 26

—with stimulant medications, 53–55

Psychological Science in the Public Interest, 160

psychological theories
—avoidance learning, 44–45
—damage hypothesis (psychological injuries of
    segregation), 2–3, 13, 175n2, 175n6
—deprivation, 6, 19, 27, 59, 62–64, 101, 141
—emotional intelligence, 7, 12, 108–13, 136–37,
    139
—experimenter bias, 7, 24, 31, 179n22
—gratification delay, 7, 11–12, 111–14, 121–37,
    201n84; and self-distraction, 197n6; skepti-
    cism about, 114–18, 126, 134–36
—grit, 8, 130
—implicit bias, 25
—interpersonal expectancy effect (Pygmalion
    effect), 7, 22–26, 29, 31, 40–43
—learned helplessness, 7, 18, 20–22, 33, 36,
    45–47, 130, 178–79n13
—learned optimism, 130
—locus of control, 29–32, 35–36, 38, 45; and
    Meier's critique of, 40
—reinforcement effects, 29, 127
—self-hatred, 3
—split-brain, 7, 11; and affect studies, 192–
    93n34; antiracist uses of, 91–102; and as-
    sumptions about gender differences, 84, 105;
    and assumptions about racial differences,
    81–82, 91–102, 108; and attributes of left
    brain, 11, 79, 86–94, 103; and attributes of
    right brain, 11, 79–82, 85–103, 105; and busi-
    ness advice, 79–80; and critiques of, 102–5;
    and enhancement of right-brain capacities,
    79–81, 88–90; and right-brain resilience,
    100–102. See also education reform
—windows of opportunity, 150, 204–5n34
Psychology: discipline of, 4, 8–9, 38–39, 52,
    113, 120; history of, 6–9, 20, 24, 120, 132–33,
    169–71; and turn away from social condi-
    tions, 113, 133, 137
Psychology of Consciousness, The (Ornstein),
    88
Psychology Today, 21, 192n34
Pygmalion (Shaw), 22–23
Pygmalion effect. See psychological theories:
    interpersonal expectancy effect
Pygmalion in the Classroom (Rosenthal and
    Jacobson), 25–26, 28, 40–43, 46

Race and Modern Science (Kuttner), 4
Racial Isolation in the Public Schools (U.S.
    Commission on Civil Rights), 26

Raising a Hyperactive Child (Stewart), 69
Ramey, Craig, 140, 151
Ramey, Sharon Landesman, 151
Rauscher, Frances, 106
Ravitch, Diane, 34, 180n39
Raz, Mical, 59
Redbook, 69
resegregation of schools, 166–67, 208n93,
    208–9n95
retardation, 19, 46, 59, 69, 71, 169. See also
    learning disability; mild mental retardation;
    minimal brain dysfunction
Rethinking the Brain (Shore), 149
rewards: and expectation of attainment, 115–
    16, 122–23, 135–36; and unattainability,
    197n6
right-brain thinking. See psychological theo-
    ries: split-brain
Ritalin, 10, 49–50, 52, 54–55, 57, 60, 62,
    67–70, 73–75, 181n40. See also stimulant
    medications
Roberts, John, 166–67, 208–9n95
Robins, Lee, 120, 127
Rosenthal, Robert, 22–26, 28, 38, 40–43
Rosenzweig, Mark R., 150, 152
Rothstein, Richard, 167, 176–77n22
Rotter, Julian, 29–30, 35, 45, 121
Ryan, William, 123, 199n52

Sagan, Carl, 80
Salovey, Peter, 109–10
Samples, Bob, 96
Saturday Review, 79, 90, 125
schizophrenia, 80, 104
Scholastic Aptitude [Assessment] Test (SAT),
    75, 170
school-to-prison pipeline, 165, 208n90
Schrag, Peter, 71–72
Schreiber, Daniel, 28
Schwartz, Gary E., 90, 192n34
Scientific American, 41, 79, 83, 89
Scientific American Mind, 135
Scott, Daryl Michael, 1
segregation, 13–14, 32, 144, 176–77n22; of
    schools, 1–4, 10, 33. See also desegregation
    of schools; resegregation of schools
self-control: and crime, 127–28; and cultivation
    of, 11–13, 130–31. See also class: and self-
    control; psychological theories: gratification
    delay

self-improvement, 8, 90, 79–81, 88–90, 108–13, 130–32, 136–37

Seligman, Martin E. P., 20–21, 43–47, 130–31, 178–79n13

Sewell, William H., 118–19

Shalit, Ruth, 76

Shatz, Carla, 140

Shaw, George Bernard, 22–23

Shepard, Samuel, Jr., 27–29

Shuey, Audrey, 4, 176n18

Silberman, Charles, 26

Skills, Technique, Academic Accomplishment, and Remediation (STAAR), 49–50

Smith, Matthew, 58

Social Cognition (Fiske and Taylor), 132

Social Learning and Clinical Psychology (Rotter), 121

socioeconomic status (SES). See class

Southern Education Report, 29

Sperry, Roger W., 82–85, 88, 97, 103

split-brain theory. See psychological theories: split-brain

Spock, Benjamin, 27, 118

Stanford-Binet test, 170

state funding for education. See government funding for education

Stell v. Savannah, 4

Stern, William, 169

Stevens, John Paul, 167, 208–9n95

Stewart, Mark A., 60–62, 69

stimulant medications, 49–57, 60–62, 67–70, 72–75, 77; as adjunct to compensatory education, 62, 72; and performance enhancement, 74–76. See also African American children: and stimulant medications; white children: and stimulant medications

Student Nonviolent Coordinating Committee, 31

Supreme Court, 1–2, 4, 36, 166

Tarnopol, Lester, 64–65

Taylor, Shelley, 132

Templeton Foundation, 159

TenHouten, Warren, 91–92, 94, 97

Terman, Lewis M., 170

testing, intelligence, 4–5, 9, 19, 31, 34, 54, 81–82, 142; and benefits of an ADD diagnosis, 75–76; and challenges to, 99–102, 109; history of, 136, 169–71, 209n1; liberal reliance on, 42–43; and reform of testing techniques,

86, 94–102, 108. See also class: and correlation with intelligence; Kaufman Assessment Battery for Children; Scholastic Aptitude [Assessment] Test; Stanford-Binet test; Wechsler Intelligence Scale for Children

testing, psychometric, 30, 67, 74, 99, 102, 119

Testing of Negro Intelligence, The (Shuey), 4, 176n18

Thompson, Andrea Lee, 95

Thorndike, Robert L., 42, 99

Time, 25, 72, 133, 136, 149

Today Show, 26

Topics in Learning and Learning Disabilities, 99

Torrance, E. Paul, 98

Tourgée, Albion, 2, 175n4

tracking within schools, 7, 10, 24, 71, 76, 98, 165–66, 170; and efforts to reduce, 28, 40–41

Trans-action, 69

transcranial magnetic stimulation (TMS), 133

trust: in tester, 121, 126; in world, 135–36

UNESCO, 5

Unheavenly City, The (Banfield), 124

Unraveling Juvenile Delinquency (Glueck and Glueck), 119

Valéry, Paul, 85

Vogel, Philip J., 83

Wall Street Journal, 145–46

Walters, Barbara, 26

War on Poverty, 9, 32, 176–77n22

Warren, Earl, 1

Washington Post, 49–50, 161

Watts, Alan, 107

Wechsler, David, 99

Wechsler Intelligence Scale for Children (WISC), 99–100

Wernicke, Paul, 82

Westinghouse study, 39, 182–83n83

white children: and ADD/ADHD, 56, 74–77; and classroom environment, 25, 33; and gifted education, 158–59, 166; and home environment, 61, 116–18; and IQ, 4–5, 46–47, 81, 95, 101–2, 146, 156; and learning disability, 71, 75–76; and left-brain dominance, 8, 82, 91–93, 95–96, 100; and locus of control, 30; and minimal brain dysfunction,

10, 59–62, 64, 66–69; and stimulant medications, 10, 52, 54, 57–62, 67–69, 71–77. *See
also* delinquency, juvenile
White House Conference on Early Childhood Development and Learning, 140–41,
148–49, 153, 156
whiteness, 7, 9, 15; and economic vulnerability, 14–15, 137; as property, 2, 175n4; and
purported discrimination against whites,
166–67; and resentment, 171; and white supremacy, 1, 13. *See also* white children
Wiesel, Torsten N., 206n53
Why Your Child Is Hyperactive (Feingold), 73

Wilson, James Q., 127
Wilson, Pete, 152
Winship, Christopher, 144
Woman's Day, 69
Wright, Richard, 31

Yerkes, Robert M., 170
yoga, 11, 88
Young Children, 150

Zen, 88–89, 107
Zero to Three (advocacy organization), 153,
203n9

STUDIES IN SOCIAL MEDICINE

Nancy M. P. King, Gail E. Henderson, and Jane Stein, eds., *Beyond Regulations: Ethics in Human Subjects Research* (1999).

Laurie Zoloth, *Health Care and the Ethics of Encounter: A Jewish Discussion of Social Justice* (1999).

Susan M. Reverby, ed., *Tuskegee's Truths: Rethinking the Tuskegee Syphilis Study* (2000).

Beatrix Hoffman, *The Wages of Sickness: The Politics of Health Insurance in Progressive America* (2000).

Margarete Sandelowski, *Devices and Desires: Gender, Technology, and American Nursing* (2000).

Keith Wailoo, *Dying in the City of the Blues: Sickle Cell Anemia and the Politics of Race and Health* (2001).

Judith Andre, *Bioethics as Practice* (2002).

Chris Feudtner, *Bittersweet: Diabetes, Insulin, and the Transformation of Illness* (2003).

Ann Folwell Stanford, *Bodies in a Broken World: Women Novelists of Color and the Politics of Medicine* (2003).

Lawrence O. Gostin, *The AIDS Pandemic: Complacency, Injustice, and Unfulfilled Expectations* (2004).

Arthur A. Daemmrich, *Pharmacopolitics: Drug Regulation in the United States and Germany* (2004).

Carl Elliott and Tod Chambers, eds., *Prozac as a Way of Life* (2004).

Steven M. Stowe, *Doctoring the South: Southern Physicians and Everyday Medicine in the Mid-Nineteenth Century* (2004).

Arleen Marcia Tuchman, *Science Has No Sex: The Life of Marie Zakrzewska, M.D.* (2006).

Michael H. Cohen, *Healing at the Borderland of Medicine and Religion* (2006).

Keith Wailoo, Julie Livingston, and Peter Guarnaccia, eds., *A Death Retold: Jesica Santillan, the Bungled Transplant, and Paradoxes of Medical Citizenship* (2006).

Michelle T. Moran, *Colonizing Leprosy: Imperialism and the Politics of Public Health in the United States* (2007).

Karey Harwood, *The Infertility Treadmill: Feminist Ethics, Personal Choice, and the Use of Reproductive Technologies* (2007).

Carla Bittel, *Mary Putnam Jacobi and the Politics of Medicine in Nineteenth-Century America* (2009).

Samuel Kelton Roberts Jr., *Infectious Fear: Politics, Disease, and the Health Effects of Segregation* (2009).

Lois Shepherd, *If That Ever Happens to Me: Making Life and Death Decisions after Terri Schiavo* (2009).

Mical Raz, *What's Wrong with the Poor? Psychiatry, Race, and the War on Poverty* (2013).

Johanna Schoen, *Abortion after* Roe (2015).

Nancy Tomes, *Remaking the American Patient: How Madison Avenue and Modern Medicine Turned Patients into Consumers* (2016).

Mara Buchbinder, Michele Rivkin-Fish, and Rebecca L. Walker, eds., *Understanding Health Inequalities and Justice: New Conversations across the Disciplines* (2016).

Muriel R. Gillick, *Old and Sick in America: The Journey through the Health Care System* (2017).

Michael E. Staub, *The Mismeasure of Minds: Debating Race and Intelligence between* Brown *and "The Bell Curve"* (2018).